Lecture Notes in Computer Science   5005

Commenced Publication in 1973
Founding and Former Series Editors:
Gerhard Goos, Juris Hartmanis, and Jan van Leeuwen

Vassilis Christophides   Martine Collard
Claudio Gutierrez (Eds.)

Semantic Web, Ontologies and Databases

VLDB Workshop, SWDB-ODBIS 2007
Vienna, Austria, September 24, 2007
Revised Selected Papers

 Springer

Volume Editors

Vassilis Christophides
University of Crete
Dept. of Computer Science
P.O. Box 2208, 71409 Herakion, Greece
E-mail: christop@ics.forth.gr

Martine Collard
INRIA, Equipe EDELWEISS
2004 route des Lucioles - BP 93, 06902 Sophia-Antipolis, France
E-mail: Martine.Collard@unice.fr

Claudio Gutierrez
Universidad de Chile
Depto. de Ciencias de la Computación
Blanco Encalada 2120, Santiago, Chile
E-mail: cgutierr@dcc.uchile.cl

Library of Congress Control Number: 2008931351

CR Subject Classification (1998): I.2.4, H.2, H.3, H.5

LNCS Sublibrary: SL 3 – Information Systems and Application, incl. Internet/Web and HCI

ISSN   0302-9743
ISBN-10   3-540-70959-2 Springer Berlin Heidelberg New York
ISBN-13   978-3-540-70959-6 Springer Berlin Heidelberg New York

Springer is a part of Springer Science+Business Media

springer.com

© Springer-Verlag Berlin Heidelberg 2008
Printed in Germany

Typesetting: Camera-ready by author, data conversion by Scientific Publishing Services, Chennai, India
Printed on acid-free paper   SPIN: 12440740   06/3180   5 4 3 2 1 0

Preface

This volume constitutes the proceedings of the joint VLDB workshop on the Semantic Web, Ontologies and Databases (SWDB-ODBIS 2007), co-located with the 33rd International Conference on Very Large Data Bases VLDB 2007. In 2007, organizers of the Semantic Web and Databases (SWDB) and the Ontologies based techniques for DataBases and Information Systems (ODBIS) workshops decided to join forces in order to further integrate these important areas of research.

Research on the semantic web and on ontologies has reached a level of maturity where the relationship with databases is becoming of paramount importance. The objective of this joint workshop was to present databases and information systems research as they relate to ontologies and the semantic web, and more broadly, to gain insight into the semantic web and ontologies as they relate to databases and information systems.

The workshop covered the foundations, methodologies and applications of these fields for databases and information systems. The papers presented focused on foundational or technological aspects as well as on research based on experience and describing industrial aspects of various topics such as semantics-aware data models and query languages; ontology-based views, mappings, transformations, and query reformulation; or storing and indexing semantic web data and schemas.

This volume includes a selection of extended versions of papers presented at the workshop. We received a total of 11 submissions, each of which was thoroughly reviewed by members of the Program Committee. We are grateful to Marcello Arenas for providing an invited talk entitled "An Extension of SPARQL for RDFS". Our sincere thanks go to the Program Committee for conducting a vigorous review process. We hope that you will find papers that are of interest to you in this volume.

March 2008

Vassilis Christophides
Martine Collard
Claudio Gutierrez

Organization

Co-chairs

Vassilis Christophides	University of Crete, Greece
Martine Collard	University of Nice-Sophia Antipolis, France
Claudio Guttierez	Catholic University of Chile, Chile

Program Committee

Bernd Amann	LIP6 - UPMC, France
Paolo Bouquet	University of Trento, Italy
Laurent Brisson	Télécom Bretagne, France
François Bry	University of Munich, Germany
Bruno Cremilleux	University of Caen, France
Philippe Cudre-Mauroux	EPFL, Switzerland
Peter W. Eklund	University of Wollongong, Australia
Carl-Christian Kanne	University of Mannheim, Germany
Bertram Ludaescher	UC Davis, USA
Michele Missikoff	Lab. for Enterprise Knowledge and Systems, IASI-CNR, Italy
Mads Nygaard	University of Trondheim, Norway
Oscar Pastor	University of Valencia, Spain
Dimitris Plexousakis	Institute of Computer Science, FORTH, Greece
Axel Polleres	DERI Galway, National University of Ireland, Galway, Ireland
Alexandra Poulovassilis	Birkbeck College, University of London, UK
Domenico Rosaci	Università Mediterranea di Reggio Calabria, Italy
Michel Scholl	CNAM, France
Vojtech Svátek	University of Economics, Prague, Czech Republic
Peter Wood	Birkbeck College, University of London, UK

Table of Contents

An Extension of SPARQL for RDFS

Marcelo Arenas[1], Claudio Gutierrez[2], and Jorge Pérez[1]

[1] Pontificia Universidad Católica de Chile
[2] Universidad de Chile

Abstract. RDF Schema (RDFS) extends RDF with a schema vocabulary with a predefined semantics. Evaluating queries which involve this vocabulary is challenging, and there is not yet consensus in the Semantic Web community on how to define a query language for RDFS. In this paper, we introduce a language for querying RDFS data. This language is obtained by extending SPARQL with *nested regular expressions* that allow to *navigate* through an RDF graph with RDFS vocabulary. This language is expressive enough to answer SPARQL queries involving RDFS vocabulary, by directly traversing the input graph.

1 Introduction

The Resource Description Framework (RDF) [16,6,14] is a data model for representing information about World Wide Web resources. The RDF specification includes a set of reserved IRIs, the RDFS vocabulary (called RDF Schema), that has a predefined semantics. This vocabulary is designed to describe special relationships between resources like typing and inheritance of classes and properties, among others features [6].

Jointly with the RDF release in 1998 as Recommendation of the W3C, the natural problem of querying RDF data was raised. Since then, several designs and implementations of RDF query languages have been proposed (see Haase et al. [12] and Furche et al. [9] for detailed comparisons of RDF query languages). In 2004, the RDF Data Access Working Group, part of the Semantic Web Activity, released a first public working draft of a query language for RDF, called SPARQL [21]. Since then, SPARQL has been rapidly adopted as the standard to query Semantic Web data. In January 2008, SPARQL became a W3C Recommendation.

The specification of SPARQL is targeted to RDF data, not including RDFS vocabulary. The reasons to follow this approach are diverse, including: (1) the lack of a standard definition of a semantics for queries under the presence of vocabulary and, hence, the lack of consensus about it; (2) the computational complexity challenges of querying in the presence of a vocabulary with a predefined semantics; and (3) practical considerations about real-life RDF data spread on the Web. These reasons explain also why most of the groups working on the definition of RDF query languages have focused in querying plain RDF data.

Nevertheless, there are several proposals to address the problem of querying RDFS data. Current practical approaches taking into account the predefined

V. Christophides et al. (Eds.): SWDB-ODBIS 2007, LNCS 5005, pp. 1–20, 2008.

semantics of the RDFS vocabulary (e.g. Harris and Gibbins [11], Broekstra et al. [7] in Sesame), roughly implement the following procedure. Given a query Q over an RDF data source G with RDFS vocabulary, the *closure* of G is computed first, that is, all the implicit information contained in G is made explicit by adding to G all the statements that are logical consequences of G. Then the query Q is evaluated over this extended data source. The theoretical formalization of such an approach was studied by Gutierrez et al. [10].

From a practical point of view, the above approach has several drawbacks. First, it is known that the size of the closure of a graph G is of quadratic order in the worst case, making the computation and storage of the closure too expensive for web-scale applications. Second, once the closure has been computed, all the queries are evaluated over a data source which can be much larger than the original one. This can be particularly inefficient for queries that must scan a large part of the input data. Third, the approach is not goal-oriented. Although in practice most queries will use just a small fragment of the RDFS vocabulary and would need only to scan a small part of the initial data, all the vocabulary and the data is considered when computing the closure.

Let us present a simple scenario that exemplifies the benefits of a goal-oriented approach. Consider an RDF data source G and a query Q that asks whether a resource A is a *sub-class of* a resource B. In its abstract syntax, RDF statements are modeled as a *subject-predicate-object* structure of the form (s, p, o), called an RDF triple. Furthermore, the keyword rdfs:subClassOf is used in RDFS to denote the sub-class relation between resources. Thus, answering Q amounts to check whether the triple $(A,$ rdfs:subClassOf$, B)$ is a logical consequence of G. The predefined semantics of RDFS states that rdfs:subClassOf is a transitive relation among resources. Then to answer Q, a goal-oriented approach should not compute the closure of the entire input graph G (which could be of quadratic order in the size of G), but instead it should just verify whether there exist resources $R_1, R_2, \ldots R_n$ such that $A = R_1$, $B = R_n$, and $(R_i,$ rdfs:subClassOf, $R_{i+1})$ is a triple in G for $i = 1, \ldots, n-1$. That is, we can answer Q by checking the existence of an rdfs:subClassOf-path from A to B in G, which takes linear time in the size of G [18].

It was shown by Muñoz el al. [18] that testing whether an RDFS triple is implied by an RDFS data source G can be done without computing the closure of G. The idea is that the RDFS deductive rules allow to determine if a triple is implied by G by essentially checking the existence of paths over G, very much like our simple example above. The good news is that these paths can be specified by using regular expressions plus some additional features. For example, to check whether $(A,$ rdfs:subClassOf$, B)$ belongs to the closure of a graph G, we already saw that it is enough to check whether there is a path from A to B in G where each edge has label rdfs:subClassOf. This observation motivates the use of extended triple patterns of the form $(A, \text{rdfs:subClassOf}^+, B)$, where rdfs:subClassOf$^+$ is the regular expression denoting paths of length at least 1 and where each edge has label rdfs:subClassOf. Thus, one can readily see that a language for navigating RDFS data would be useful for obtaining the answer of queries considering the predefined semantics of the RDFS vocabulary.

Driven by this motivation, in this paper we introduce a language that extends SPARQL with navigational capabilities. The resulting language turns out to be expressive enough to capture the deductive rules of RDFS. Thus, we can obtain the RDFS evaluation of an important fragment of SPARQL by navigating directly the input RDFS data source, without computing the closure.

This idea can be developed at several levels. We first consider a navigational language that includes regular expressions and takes advantage of the special features of RDF. Paths defined by regular expressions has been widely used in graph databases [17,3], and recently, have been also proposed in the RDF context [1,4,2,15,5]. We show that although paths defined in terms of regular expressions are useful, regular expressions alone are not enough to obtain the RDFS evaluation of some queries by simply navigating RDF data. Thus, we enrich regular expressions by borrowing the notion of *branching* from XPath [8], to obtain what we call *nested regular expressions*. Nested regular expressions are enough for our purposes and, furthermore, they provide an interesting extra expressive power to define complex path queries over RDF data with RDFS vocabulary.

Organization of the paper. In Section 2, we present a summary of the basics of RDF, RDFS, and SPARQL, based on Muñoz et al. [18] and Pérez et al. [20]. Section 3 is the core part of the paper, and introduces our proposal for a navigational language for RDF. We first discuss the related work on navigating RDF in Section 3.1. In Section 3.2, we introduce a first language for navigating RDF graphs based on regular expressions, and we discuss why regular expressions alone are not enough for our purposes. Section 3.3 presents the language of nested regular expressions, and shows how these expressions can be used to obtain the RDFS evaluation of SPARQL patterns. In Section 3.4, we give some examples of the extra expressive power of nested regular expressions, showing the usefulness of the language to extract complex path relations from RDF graphs. Finally, Section 4 presents some conclusions.

2 RDFS and SPARQL

In this section, we present the algebraic formalization of the core fragment of SPARQL over RDF graphs introduced in [20], and then we extend this formalization to RDFS graphs. But before doing that, we introduce some notions related to RDF and the core fragment of RDFS.

2.1 The RDF Data Model

RDF is a graph data format for representing information in the Web. An RDF statement is a *subject-predicate-object* structure, called an RDF *triple*, intended to describe resources and properties of those resources. For the sake of simplicity, we assume that RDF data is composed only by elements from an infinite set U

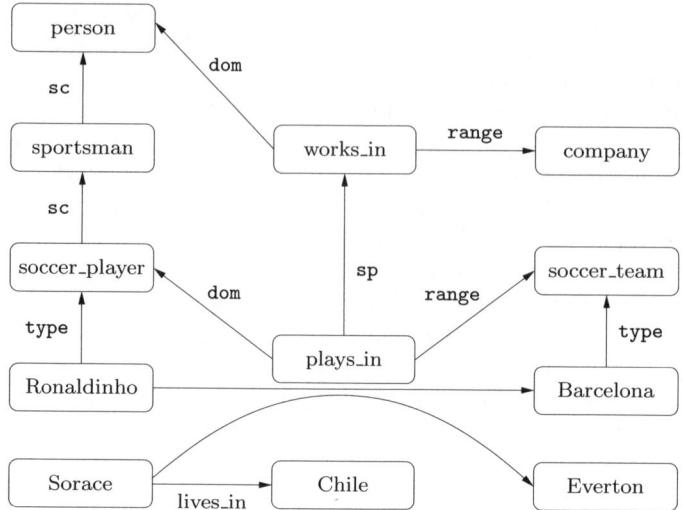

Fig. 1. An RDF graph storing information about soccer players

of IRIs[1]. More formally, an RDF triple is a tuple $(s, p, o) \in U \times U \times U$, where s is the *subject*, p the *predicate* and o the *object*. An RDF graph (or RDF data source) is a finite set of RDF triples.

Figure 1 shows an RDF graph that stores information about soccer players. In this figure, a triple (s, p, o) is depicted as an arc $s \xrightarrow{p} o$, that is, s and o are represented as nodes and p is represented as an arc label. For example, (Sorace, lives_in, Chile) is a triple in the RDF graph in Figure 1. Notice that, an RDF graph is not a standard labeled graph as its set of labels may have a nonempty intersection with its set of nodes. For instance, consider triples (Ronaldinho, plays_in, Barcelona) and (plays_in, sp, works_in) in the RDF graph in Figure 1. In this example, plays_in is the predicate of the first triple and the subject of the second one, and thus, acts simultaneously as a node and an edge label.

The RDF specification includes a set of reserved IRIs (reserved elements from U) with predefined semantics, the RDFS vocabulary (RDF Schema [6]). This set of reserved words is designed to deal with inheritance of classes and properties, as well as typing, among other features [6]. In this paper, we consider the subset of the RDFS vocabulary composed by the special IRIs rdfs:subClassOf, rdfs:subPropertyOf, rdfs:range, rdfs:domain and rdf:type, which are denoted by sc, sp, range, dom and type, respectively. The RDF graph in Figure 1 uses these keywords to relate resources. For instance, the graph contains triple (sportsman, sc, person), thus stating that sportsman is a *sub-class of* person.

The fragment of RDFS consisting of the keywords sc, sp, range, dom and type was considered in [18]. In that paper, the authors provide a formal semantics

---

[1] In this paper, we do not consider anonymous resources called blank nodes in the RDF data model, that is, our study focus on *ground* RDF graphs. We neither make a special distinction between IRIs and Literals.

for it, and also show it to be well-behaved as the remaining RDFS vocabulary does not interfere with the semantics of this fragment. This together with some other results from [18] provide strong theoretical and practical evidence for the importance of this fragment. In this paper, we consider the keywords sc, sp, range, dom and type, and we use the semantics for them from [18], instead of using the full RDFS semantics (these two were shown to be equivalent in [18]).

For the sake of simplicity, we do not include here the model theoretical semantics for RDFS from [18], and we only present the system of rules from [18] that was proved to be equivalent to the model theoretical semantics (that is, was proved to be sound and complete for the inference problem for RDFS in the presence of sc, sp, range, dom and type). Table 1 shows the inference system for the fragment of RDFS considered in this paper. Next we formalize the notion of deduction for this system of inference rules. In every rule, letters $\mathcal{A}$, $\mathcal{B}$, $\mathcal{C}$, $\mathcal{X}$, and $\mathcal{Y}$, stand for *variables* to be replaced by actual terms. More formally, an *instantiation* of a rule is a replacement of the variables occurring in the triples of the rule by elements of U. An *application* of a rule to a graph G is defined as follows. Given a rule r, if there is an instantiation $\frac{R}{R'}$ of r such that $R \subseteq G$, then the graph $G' = G \cup R'$ is the result of an application of r to G. Finally, the *closure* of an RDF graph G, denoted by cl(G), is defined as the graph obtained from G by successively applying the rules in Table 1 until the graph does not change.

Example 1. Consider the RDF graph in Figure 1. By applying the rule (1b) to (Ronaldinho, plays_in, Barcelona) and (plays_in, sp, works_in), we obtain that (Ronaldinho, works_in, Barcelona) is in the closure of the graph. Moreover, by applying the rule (3b) to this last triple and (works_in, range, company), we obtain that (Barcelona, type, company) is also in the closure of the graph. Figure 2 shows the complete closure of the RDF graph in Figure 1. The solid lines in Figure 2 represent the triples in the original graph, and the dashed lines the additional triples in the closure. □

In [18], it was shown that if the number of triples in G is n, then the closure cl(G) could have, in the worst case, $\Omega(n^2)$ triples.

Table 1.

1. *Subproperty:*

    (a) $\dfrac{(\mathcal{A},\text{sp},\mathcal{B})\ \ (\mathcal{B},\text{sp},\mathcal{C})}{(\mathcal{A},\text{sp},\mathcal{C})}$

    (b) $\dfrac{(\mathcal{A},\text{sp},\mathcal{B})\ \ (\mathcal{X},\mathcal{A},\mathcal{Y})}{(\mathcal{X},\mathcal{B},\mathcal{Y})}$

2. *Subclass:*

    (a) $\dfrac{(\mathcal{A},\text{sc},\mathcal{B})\ \ (\mathcal{B},\text{sc},\mathcal{C})}{(\mathcal{A},\text{sc},\mathcal{C})}$

    (b) $\dfrac{(\mathcal{A},\text{sc},\mathcal{B})\ \ (\mathcal{X},\text{type},\mathcal{A})}{(\mathcal{X},\text{type},\mathcal{B})}$

3. *Typing:*

    (a) $\dfrac{(\mathcal{A},\text{dom},\mathcal{B})\ \ (\mathcal{X},\mathcal{A},\mathcal{Y})}{(\mathcal{X},\text{type},\mathcal{B})}$

    (b) $\dfrac{(\mathcal{A},\text{range},\mathcal{B})\ \ (\mathcal{X},\mathcal{A},\mathcal{Y})}{(\mathcal{Y},\text{type},\mathcal{B})}$

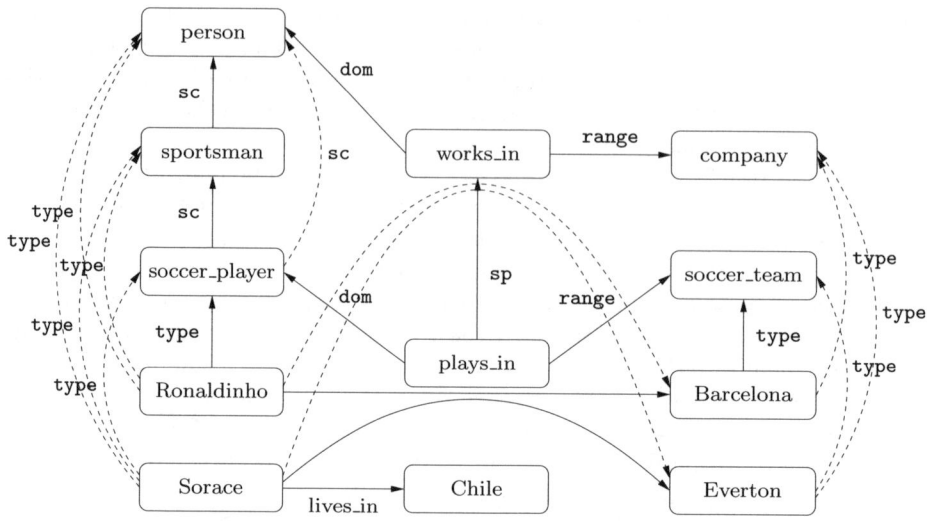

Fig. 2. The closure of the RDF graph in Figure 1

2.2   SPARQL

SPARQL is essentially a graph-matching query language. A SPARQL query is of the form $H \leftarrow B$. The *body B* of the query, is a complex RDF graph pattern expression that may include RDF triples with variables, conjunctions, disjunctions, optional parts and constraints over the values of the variables. The *head H* of the query, is an expression that indicates how to construct the answer to the query. The evaluation of a query Q against an RDF graph G is done in two steps: the body of Q is matched against G to obtain a set of bindings for the variables in the body, and then using the information on the head of Q, these bindings are processed applying classical relational operators (projection, distinct, etc.) to produce the answer to the query. This answer can have different forms, e.g. a yes/no answer, a table of values, or a new RDF graph. In this paper, we concentrate on the body of SPARQL queries, i.e. in the graph pattern matching facility.

Assume the existence of an infinite set V of variables disjoint from U. A SPARQL graph pattern is defined recursively as follows [20]:

1. A tuple from $(U \cup V) \times (U \cup V) \times (U \cup V)$ is a graph pattern (a *triple pattern*).
2. If P_1 and P_2 are graph patterns, then expressions $(P_1 \text{ AND } P_2)$, $(P_1 \text{ OPT } P_2)$, and $(P_1 \text{ UNION } P_2)$ are graph patterns.
3. If P is a graph pattern and R is a SPARQL *built-in* condition, then the expression $(P \text{ FILTER } R)$ is a graph pattern.

A SPARQL *built-in* condition is a Boolean combination of terms constructed by using the equality $(=)$ among elements in $U \cup V$ and constant, and the unary predicate bound($\cdot$) over variables.

To define the semantics of SPARQL graph patterns, we need to introduce some terminology. A *mapping* μ from V to U is a partial function $\mu : V \rightarrow U$. Slightly abusing notation, for a triple pattern t we denote by $\mu(t)$ the triple obtained by replacing the variables in t according to μ. The domain of μ, denoted by $\text{dom}(\mu)$, is the subset of V where μ is defined. Two mappings μ_1 and μ_2 are *compatible* if for every $x \in \text{dom}(\mu_1) \cap \text{dom}(\mu_2)$, it is the case that $\mu_1(x) = \mu_2(x)$, i.e. when $\mu_1 \cup \mu_2$ is also a mapping. Intuitively, μ_1 and μ_2 are compatibles if μ_1 *can be extended* with μ_2 to obtain a new mapping, and vice versa. Note that two mappings with disjoint domains are always compatible, and that the empty mapping $\mu_\emptyset$ (i.e. the mapping with empty domain) is compatible with any other mapping.

Let Ω_1 and Ω_2 be sets of mappings. We define the join of, the union of and the difference between Ω_1 and Ω_2 as:

$$\Omega_1 \bowtie \Omega_2 = \{\mu_1 \cup \mu_2 \mid \mu_1 \in \Omega_1, \mu_2 \in \Omega_2 \text{ and } \mu_1, \mu_2 \text{ are compatible mappings}\},$$
$$\Omega_1 \cup \Omega_2 = \{\mu \mid \mu \in \Omega_1 \text{ or } \mu \in \Omega_2\},$$
$$\Omega_1 \smallsetminus \Omega_2 = \{\mu \in \Omega_1 \mid \text{ for all } \mu' \in \Omega_2, \mu \text{ and } \mu' \text{ are not compatible}\}.$$

Based on the previous operators, we define the left outer-join as:

$$\Omega_1 \bowtie\kern-1.2em\rule[0.1ex]{0.3em}{0.12ex}\kern0.6em \Omega_2 = (\Omega_1 \bowtie \Omega_2) \cup (\Omega_1 \smallsetminus \Omega_2).$$

Intuitively, $\Omega_1 \bowtie \Omega_2$ is the set of mappings that result from extending mappings in Ω_1 with their compatible mappings in Ω_2, and $\Omega_1 \smallsetminus \Omega_2$ is the set of mappings in Ω_1 that cannot be extended with any mapping in Ω_2. The operation $\Omega_1 \cup \Omega_2$ is the usual set theoretical union. A mapping μ is in $\Omega_1 \bowtie\kern-1.2em\rule[0.1ex]{0.3em}{0.12ex}\kern0.6em \Omega_2$ if it is the extension of a mapping of Ω_1 with a compatible mapping of Ω_2, or if it belongs to Ω_1 and cannot be extended with any mapping of Ω_2. These operations resemble relational algebra operations over sets of mappings (partial functions).

We are ready to define the semantics of graph pattern expressions as a function that takes a pattern expression and returns a set of mappings. The *evaluation* of a graph pattern over an RDF graph G, denoted by $[\![\cdot]\!]_G$, is defined recursively as follows:

- $[\![t]\!]_G = \{\mu \mid \text{dom}(\mu) = \text{var}(t) \text{ and } \mu(t) \in G\}$, where $\text{var}(t)$ is the set of variables occurring in t.
- $[\![(P_1 \text{ AND } P_2)]\!]_G = [\![P_1]\!]_G \bowtie [\![P_2]\!]_G$.
- $[\![(P_1 \text{ UNION } P_2)]\!]_G = [\![P_1]\!]_G \cup [\![P_2]\!]_G$.
- $[\![(P_1 \text{ OPT } P_2)]\!]_G = [\![P_1]\!]_G \bowtie\kern-1.2em\rule[0.1ex]{0.3em}{0.12ex}\kern0.6em [\![P_2]\!]_G$.

The idea behind the OPT operator is to allow for *optional matching* of patterns. Consider pattern expression $(P_1 \text{ OPT } P_2)$ and let μ_1 be a mapping in $[\![P_1]\!]_G$. If there exists a mapping $\mu_2 \in [\![P_2]\!]_G$ such that μ_1 and μ_2 are compatible, then $\mu_1 \cup \mu_2$ belongs to $[\![(P_1 \text{ OPT } P_2)]\!]_G$. But if no such a mapping μ_2 exists, then μ_1 belongs to $[\![(P_1 \text{ OPT } P_2)]\!]_G$. Thus, operator OPT allows information to be added to a mapping μ if the information is available, instead of just rejecting μ whenever some part of the pattern does not match.

The semantics of FILTER expressions goes as follows. Given a mapping μ and a built-in condition R, we say that μ satisfies R, denoted by $\mu \models R$, if:

- R is bound(?X) and ?$X \in dom(\mu)$;
- R is ?$X = c$, ?$X \in dom(\mu)$ and $\mu(?X) = c$;
- R is ?$X =$?Y, ?$X \in dom(\mu)$, ?$Y \in dom(\mu)$ and $\mu(?X) = \mu(?Y)$;
- R is $(\neg R_1)$, R_1 is a built-in condition, and it is not the case that $\mu \models R_1$;
- R is $(R_1 \vee R_2)$, R_1 and R_2 are built-in conditions, and $\mu \models R_1$ or $\mu \models R_2$;
- R is $(R_1 \wedge R_2)$, R_1 and R_2 are built-in conditions, $\mu \models R_1$ and $\mu \models R_2$.

Then $[\![(P \text{ FILTER } R)]\!]_G = \{\mu \in [\![P]\!]_G \mid \mu \models R\}$, that is, $[\![(P \text{ FILTER } R)]\!]_G$ is the set of mappings in $[\![P]\!]_G$ that satisfy R.

It was shown in [20], among other algebraic properties, that AND and UNION are associative and commutative, thus permitting us to avoid parenthesis when writing sequences of either AND operators or UNION operators.

In the rest of the paper, we usually represent sets of mappings as tables where each row represents a mapping in the set. We label every row with the name of a mapping, and every column with the name of a variable. If a mapping is not defined for some variable, then we simply leave empty the corresponding position. For instance, the table:

	?X	?Y	?Z	?V	?W
μ_1 :	a	b			
μ_2 :		c			d
μ_3 :			e		

represents the set $\Omega = \{\mu_1, \mu_2, \mu_3\}$, where

- $dom(\mu_1) = \{?X, ?Y\}$, $\mu_1(?X) = a$ and $\mu_1(?Y) = b$;
- $dom(\mu_2) = \{?Y, ?W\}$, $\mu_2(?Y) = c$ and $\mu_2(?W) = d$;
- $dom(\mu_3) = \{?Z\}$ and $\mu_3(?Z) = e$.

We sometimes write $\{\{?X \rightarrow a, ?Y \rightarrow b\}, \{?Y \rightarrow c, ?W \rightarrow d\}, \{?Z \rightarrow e\}\}$ for the above set of mappings.

Example 2. Let G be the RDF graph shown in Figure 1, and consider SPARQL graph pattern $P_1 = ((?X, \text{plays\_in}, ?T) \text{ AND } (?X, \text{lives\_in}, ?C))$. Intuitively, P_1 retrieves the list of soccer players in G, including the teams where they play in and the countries where they live in. Thus, we have:

$$[\![P_1]\!]_G = \begin{array}{|c|c|c|} \hline ?X & ?T & ?C \\ \hline \text{Sorace} & \text{Everton} & \text{Chile} \\ \hline \end{array}$$

Notice that in this case we have not obtained any information about Ronaldinho, since in the graph there is not data about the country where Ronaldinho lives in. Consider now the pattern $P_2 = ((?X, \text{plays\_in}, ?T) \text{ OPT } (?X, \text{lives\_in}, ?C))$. Intuitively, P_2 retrieves the list of soccer players in G, including the teams where they play in and the countries where they live in. But, as opposed to P_1, pattern

P_2 does not fail if the information about the country where a soccer player lives in is missing. In this case, we have:

$$[\![P_2]\!]_G =$$

?X	?T	?C
Sorace	Everton	Chile
Ronaldinho	Barcelona	

$\square$

2.3   The Semantics of SPARQL over RDFS

SPARQL follows a *subgraph-matching* approach, and thus, a SPARQL query treats RDFS vocabulary without considering its predefined semantics. For instance, let G be the RDF graph shown in Figure 1, and consider the graph pattern $P = (?X, \text{works\_in}, ?C)$. Note that, although the triples (Ronaldinho, works_in, Barcelona) and (Sorace, works_in, Everton) can be deduced from G, we obtain the empty set as the result of evaluating P over G (that is, $[\![P]\!]_G = \emptyset$) as there is no triple in G with works_in in the predicate position.

We are interested in defining the semantics of SPARQL over RDFS, that is, taking into account not only the explicit RDF triples of a graph G, but also the triples that can be derived from G according to the semantics of RDFS. The most direct way of defining such a semantics is by considering not the original graph but its closure. The following definition formalizes this notion.

Definition 1 (RDFS evaluation). *Given a SPARQL graph pattern P, the RDFS evaluation of P over G, denoted by $[\![P]\!]_G^{\text{rdfs}}$, is defined as the set of mappings $[\![P]\!]_{\text{cl}(G)}$, that is, as the evaluation of P over the closure of G.*

Example 3. Let G be the RDF graph shown in Figure 1, and consider the graph pattern expression:

$$P = ((?X, \text{type}, \text{person}) \text{ AND } (?X, \text{lives\_in}, \text{Chile}) \text{ AND } (?X, \text{works\_in}, ?C)),$$

intended to retrieve the list of people in G (resources of type person) that lives in Chile, and the companies where they work in. The evaluation of P over G results in the empty set, since both $[\![(?X, \text{type}, \text{person})]\!]_G$ and $[\![(?X, \text{works\_in}, ?C)]\!]_G$ are empty. On the other hand, the RDFS evaluation of P over G contains the following tuples:

$$[\![P]\!]_G^{\text{rdfs}} = [\![P]\!]_{\text{cl}(G)} =$$

?X	?C
Sorace	Everton

$\square$

It should be noticed that in Definition 1, we do not provide a procedure for evaluating SPARQL over RDFS. In fact, as we have mentioned before, a direct implementation of this definition leads to an inefficient procedure for evaluating SPARQL queries, as it requires a pre-calculation of the closure of the input graph.

3 Navigational RDF Languages

Our main goal is to define a query language that allows to obtain the RDFS evaluation of a pattern directly from an RDF graph, without computing the entire closure of the graph. We have provided some evidence that a language for navigating RDF graphs could be useful in achieving our goal. In this section, we define such a language for *navigating* RDF graphs, providing a formal syntax and semantics. Our language uses, as usual for graph query languages [17,3], regular expressions to define paths on graph structures, but taking advantage of the special features of RDF graphs. More precisely, we start by introducing in Section 3.2 a language that extends SPARQL with regular expressions. Although regular expressions capture in some cases the semantics of RDFS, we show in Section 3.2 that regular expressions alone are not enough to obtain the RDFS evaluation of some queries. Thus, we show in Section 3.3 how to extend regular expressions by borrowing the notion of *branching* from XPath [8], and we explain why this enriched language is enough for our purposes. Finally, we show in Section 3.4 that the enriched language provides some other interesting features that give extra expressiveness to the language, and that deserve further investigation. But before doing all this, we briefly review in Section 3.1 some of the related work on navigating RDF.

3.1 Related Work

The idea of having a language to navigate through an RDF graph is not new. In fact, several languages have been proposed in the literature [1,4,2,15,5]. Nevertheless, none of these languages is motivated by the necessity to evaluate queries over RDFS, and none of them is comparable in expressiveness with the language proposed in this paper. Kochut et al. [15] propose a language called SPARQLeR as an extension of SPARQL. This language allows to extract semantic associations between RDF resources by considering paths in the input graph. SPARQLeR works with path variables intended to represent a sequence of resources in a path between two nodes in the input graph. A SPARQLeR query can also put restrictions over those paths by checking whether they conform to a regular expression. With the same motivation of extracting semantic associations from RDF graphs, Anyanwu et al. [5] propose a language called SPARQ2L. SPARQ2L extends SPARQL by allowing path variables and path constraints. For example, some SPARQ2L constraints are based on the presence (or absence) of some nodes or edges, the length of the retrieved paths, and on some structural properties of these paths. In [5], the authors also investigate the implementation of a query evaluation mechanism for SPARQ2L with emphasis in some secondary memory issues.

The language PSPARQL was proposed by Alkhateeb et al. in [2]. PSPARQL is an extension of SPARQL obtained by allowing regular expressions in the predicate position of triple patterns. Thus, this language can be used to obtain pair of nodes that are connected by a path whose labeling conforms to a regular expression. PSPARQL also allows variables inside regular expressions, thus permitting to retrieve data *along* the traversed paths. In [2], the authors propose a formal semantics for PSPARQL, and also study some theoretical aspects of this language such as the complexity of query evaluation. VERSA [19] and RxPath [22] are proposals motivated by XPath with emphasis on some implementation issues.

3.2   Navigating RDF through Regular Expressions

Navigating graphs is done usually by using an operator *next*, which allows to move from one node to an adjacent one in a graph. In our setting, we have RDF "graphs", which are sets of triples, not classical graphs [13]. In particular, instead of classical edges (pair of nodes), we have directed triples of nodes (*hyperedges*). Hence, a language for navigating RDF graphs should be able to deal with this type of objects. The language introduced in this paper deals with this problem by using three different navigation axes, which are shown in Figure 3 (together with their inverses).

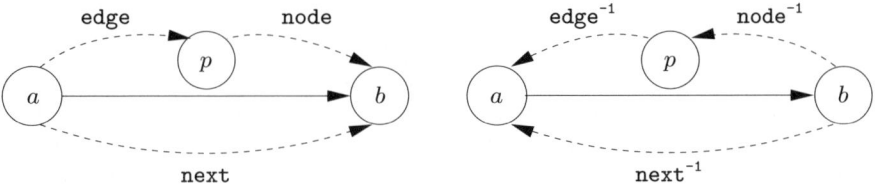

Fig. 3. Forward and backward axes for an RDF triple (a, p, b)

A navigation axis allows moving one step forward (or backward) in an RDF graph. Thus, a sequence of these axes defines a path in an RDF graph, and one can use classical regular expressions over these axes to define a set of paths that can be used in a query. More precisely, the following grammar defines the regular expressions in our language:

$$exp \quad := \quad axis \mid axis::a \ (a \in U) \mid exp/exp \mid exp|exp \mid exp^* \qquad (1)$$

where $axis \in \{\texttt{self}, \texttt{next}, \texttt{next}^{-1}, \texttt{edge}, \texttt{edge}^{-1}, \texttt{node}, \texttt{node}^{-1}\}$. The additional axis $\texttt{self}$ is not used to navigate, but instead to test the label of a specific node in a path. We call *regular path expressions* to expressions generated by (1).

Before introducing the formal semantics of regular path expressions, we give some intuition about how these expressions are evaluated in an RDF graph. The most natural navigation axis is $\texttt{next}::a$, with a an arbitrary element from U. Given an RDF graph G, the expression $\texttt{next}::a$ is interpreted as the *a-neighbor* relation in G, that is, the pairs of nodes (x, y) such that $(x, a, y) \in G$. Given that in the RDF data model a node can also be the label of an edge, the language allows to navigate from a node to one of its leaving edges by using the $\texttt{edge}$ axis. More formally, the interpretation of $\texttt{edge}::a$ is the pairs of nodes (x, y) such that $(x, y, a) \in G$. We formally define the evaluation of a regular path expression p in a graph G as a binary relation $[\![p]\!]_G$, denoting the pairs of nodes (x, y) such that

Table 2. Formal semantics of regular path expressions

$$[\![\mathtt{self}]\!]_G = \{(x,x) \mid x \in \mathrm{voc}(G)\}$$
$$[\![\mathtt{self::}a]\!]_G = \{(a,a)\}$$

$$[\![\mathtt{next}]\!]_G = \{(x,y) \mid \text{there exists } z \text{ s.t. } (x,z,y) \in G\}$$
$$[\![\mathtt{next::}a]\!]_G = \{(x,y) \mid (x,a,y) \in G\}$$
$$[\![\mathtt{edge}]\!]_G = \{(x,y) \mid \text{there exists } z \text{ s.t. } (x,y,z) \in G\}$$
$$[\![\mathtt{edge::}a]\!]_G = \{(x,y) \mid (x,y,a) \in G\}$$
$$[\![\mathtt{node}]\!]_G = \{(x,y) \mid \text{there exists } z \text{ s.t. } (z,x,y) \in G\}$$
$$[\![\mathtt{node::}a]\!]_G = \{(x,y) \mid (a,x,y) \in G\}$$

$$[\![\mathrm{axis}^{-1}]\!]_G = \{(x,y) \mid (y,x) \in [\![\mathrm{axis}]\!]_G\} \quad \text{with axis} \in \{\mathtt{next},\mathtt{node},\mathtt{edge}\}$$
$$[\![\mathrm{axis}^{-1}::a]\!]_G = \{(x,y) \mid (y,x) \in [\![\mathrm{axis::}a]\!]_G\} \quad \text{with axis} \in \{\mathtt{next},\mathtt{node},\mathtt{edge}\}$$

$$[\![exp_1/exp_2]\!]_G = \{(x,y) \mid \text{there exists } z \text{ s.t. } (x,z) \in [\![exp_1]\!]_G \text{ and } (z,y) \in [\![exp_2]\!]_G\}$$
$$[\![exp_1 \mid exp_2]\!]_G = [\![exp_1]\!]_G \cup [\![exp_2]\!]_G$$
$$[\![exp^*]\!]_G = [\![\mathtt{self}]\!]_G \cup [\![exp]\!]_G \cup [\![exp/exp]\!]_G \cup [\![exp/exp/exp]\!]_G \cup \cdots$$

y is reachable from x in G by following a path whose labels are in the language defined by p. The formal semantics of the language is shown in Table 2. In this table, G is an RDF graph, $a \in U$, $\mathrm{voc}(G)$ is the set of all the elements from U that are mentioned in G, and exp, exp_1, exp_2 are regular path expressions.

Example 4. Consider an RDF graph G storing information about transportation services between cities. A triple (C_1, tc, C_2) in the graph indicates that there is a direct way of traveling from C_1 to C_2 by using the transportation company tc.

If we assume that G does not mention any of the RDFS keywords, then the expression:

$$(\mathtt{next::}\mathrm{KoreanAir})^+ \mid (\mathtt{next::}\mathrm{AirFrance})^+$$

defines the pairs of cities (C_1, C_2) in G such that there is a way of flying from C_1 to C_2 in either KoreanAir or AirFrance. Moreover, by using axis $\mathtt{self}$, we can test for a stop in a specific city. For example, the expression:

$$(\mathtt{next::}\mathrm{KoreanAir})^+/\mathtt{self::}\mathrm{Paris}/(\mathtt{next::}\mathrm{KoreanAir})^+$$

defines the pairs of cities (C_1, C_2) such that there is a way of flying from C_1 to C_2 with KoreanAir with a stop in Paris. $\square$

Once regular path expressions have been defined, the natural next step is to extend the syntax of SPARQL to allow them in triple patterns. A *regular path triple* is a tuple of the form $t = (x, exp, y)$, where $x, y \in U \cup V$ and exp is a regular path expression. Then the evaluation of a regular path triple $t = (?X, exp, ?Y)$ over an RDF graph G is defined as the following set of mappings:

$$[\![t]\!]_G = \{\mu \mid \mathrm{dom}(\mu) = \{?X, ?Y\} \text{ and } (\mu(?X), \mu(?Y)) \in [\![exp]\!]_G\}.$$

Similarly, the evaluation of a regular path triple $t = (?X, exp, a)$ over an RDF graph G, where $a \in U$, is defined as $\{\mu \mid \text{dom}(\mu) = \{?X\} \text{ and } (\mu(?X), a) \in [\![exp]\!]_G\}$, and likewise for $(a, exp, ?X)$ and (a, exp, b) with $b \in U$.

We call *regular* SPARQL (or just rSPARQL) to SPARQL extended with regular path triples. The semantics of rSPARQL patterns is defined recursively as in Section 2, but considering the special semantics of regular path triples. The following example shows that rSPARQL is useful to represent RDFS deductions.

Example 5. Let G be the RDF graph in Figure 1, and assume that we want to obtain the *type* information of Ronaldinho. This information can be obtained by computing the RDFS evaluation of the pattern (Ronaldinho, type, $?C$). By simply inspecting the closure of G in Figure 2, we obtain that:

$$[\![(\text{Ronaldinho}, \texttt{type}, ?C)]\!]_G^{\text{rdfs}} = \begin{array}{|c|} \hline ?C \\ \hline \text{soccer\_player} \\ \hline \text{sportsman} \\ \hline \text{person} \\ \hline \end{array}$$

However, if we directly evaluate this pattern over G we obtain a single mapping:

$$[\![(\text{Ronaldinho}, \texttt{type}, ?C)]\!]_G = \begin{array}{|c|} \hline ?C \\ \hline \text{soccer\_player} \\ \hline \end{array}$$

Consider now the rSPARQL pattern:

$$P = (\text{Ronaldinho}, \texttt{next::type/(next::sc)}^*, ?C).$$

The regular path expression $\texttt{next::type/(next::sc)}^*$ is intended to obtain the pairs of nodes such that, there is a path between them that has type as its first label followed by zero or more labels sc. When evaluating this expression in G, we obtain the set of pairs {(Ronaldinho, soccer_player), (Ronaldinho, sportsman), (Ronaldinho, person), (Barcelona, soccer_team)}. Thus, the evaluation of P results in the set of mappings:

$$[\![P]\!]_G = \begin{array}{|c|} \hline ?C \\ \hline \text{soccer\_player} \\ \hline \text{sportsman} \\ \hline \text{person} \\ \hline \end{array}$$

In this case, pattern P is enough to obtain the type information of Ronaldinho in G according to the RDFS semantics, that is,

$$[\![(\text{Ronaldinho}, \texttt{type}, ?C)]\!]_G^{\text{rdfs}} = [\![(\text{Ronaldinho}, \texttt{next::type/(next::sc)}^*, ?C)]\!]_G.$$

Although the expression $\texttt{next::type/(next::sc)}^*$ is enough to obtain the type information for Ronaldinho in G, it cannot be used in general to obtain the type information of a resource. For instance, in the same graph, assume that we want to obtain the type information of Everton. In this case, if we evaluate the pattern (Everton, `next::type`/(`next::sc`)*, ?C) over G, we obtain the empty set. Consider now the rSPARQL pattern

$$Q = (\text{Everton}, \texttt{node}^{-1}/(\texttt{next::sp})^*/\texttt{next::range}, ?C).$$

With the expression `node`$^{-1}$/(`next::sp`)*/`next::range`, we follow a path that first navigates from a node to one of its incoming edges by using `node`$^{-1}$, and then continues with zero or more `sp` edges and a final `range` edge. The evaluation of this expression in G results in the set {(Everton, soccer_team), (Everton, company), (Barcelona, soccer_team), (Barcelona, company)}. Thus, the evaluation of Q in G is the set of mappings:

$$\llbracket Q \rrbracket_G = \begin{array}{|c|} \hline ?C \\ \hline \text{soccer\_team} \\ \hline \text{company} \\ \hline \end{array}$$

By looking at the closure of G in Figure 2, we see that pattern Q obtains exactly the type information of Everton in G, that is, $\llbracket(\text{Everton}, \texttt{type}, ?C)\rrbracket_G^{\text{rdfs}} = \llbracket Q \rrbracket_G$.

$\square$

The previous example shows the benefits of having regular path expressions to obtain the RDFS evaluation of a pattern P over an RDF graph G just by navigating G. We are interested in whether this can be done in general for every SPARQL pattern. More formally, we are interested in the following problem:

Given a SPARQL pattern P, is there an rSPARQL pattern Q such that for every RDF graph G, it holds that

$$\llbracket P \rrbracket_G^{\text{rdfs}} = \llbracket Q \rrbracket_G?$$

Unfortunately, the answer to this question is negative for some SPARQL patterns. Let us show this failure with an example. Assume that we want to obtain the RDFS evaluation of pattern $P = (?X, \text{works\_in}, ?Y)$ in an RDF graph G. This can be done by first finding all the properties p that are sub-properties of works_in, and then finding all the resources a and b such that (a, p, b) is a triple in G. A way to answer P by navigating the graph would be to find the pairs of nodes (a, b) such that there is a path from a to b that: (1) goes from a to one of its leaving edges, then (2) follows a sequence of zero or more `sp` edges until it reaches a works_in edge, and finally (3) returns to the initial edge and moves forward to b. If such a path exists, then it is clear that $(a, \text{works\_in}, b)$ can be deduced from the graph. The following is a natural attempt to obtain the described path with a regular path expression:

$$\texttt{edge}/(\texttt{next::sp})^*/\texttt{self::works\_in}/(\texttt{next}^{-1}\texttt{::sp})^*/\texttt{node}.$$

The problem with the above expression is that, when the path returns from works_in, no information about the path used to reach works_in has been stored. Thus, there is no way to know what was the initial edge. In fact, if we evaluate the pattern $Q = (?X, \texttt{edge}/(\texttt{next::sp})^*/\texttt{self::works\_in}/(\texttt{next}^{-1}\texttt{::sp})^*/\texttt{node}, ?Y)$ over the graph G in Figure 1, we obtain the set of mappings:

$$\llbracket Q \rrbracket_G =$$

?X	?Y
Ronaldinho	Barcelona
Ronaldinho	Everton
Sorace	Barcelona
Sorace	Everton

By simply inspecting the closure of G in Figure 2, we obtain that:

$$\llbracket P \rrbracket_G^{\text{rdfs}} =$$

?X	?Y
Ronaldinho	Barcelona
Sorace	Everton

and, thus, we have that Q is not the right representation of P according to the RDFS semantics, since $\llbracket P \rrbracket_G^{\text{rdfs}} \neq \llbracket Q \rrbracket_G$.

In general, it can be shown that there is no rSPARQL triple pattern Q such that for every RDF graph G, it holds that $\llbracket (?X, \text{works\_in}, ?Y) \rrbracket_G^{\text{rdfs}} = \llbracket Q \rrbracket_G$. It is worth mentioning that this failure persists for a general rSPARQL pattern Q, that is, if Q is allowed to use all the expressive power of SPARQL patterns (it can use operators AND, UNION, OPT and FILTER) plus regular path expressions in triple patterns.

3.3 Navigating RDF through Nested Regular Expressions

We have seen that regular path expressions are not enough to obtain the RDFS evaluation of a graph pattern. In this section, we introduce a language that extends regular path expressions with a *nesting operator*. Nested expressions can be used to test for the existence of certain paths starting at any axis of a regular path expression. We will see that this feature is crucial in obtaining the RDFS evaluation of SPARQL patterns by directly traversing RDF graphs.

The syntax of nested regular expressions is defined by the following grammar:

$$exp \ := \ \text{axis} \ | \ \text{axis::}a \ (a \in U) \ | \ \text{axis::}[exp] \ | \ exp/exp \ | \ exp|exp \ | \ exp^* \quad (2)$$

where axis $\in \{\texttt{self}, \texttt{next}, \texttt{next}^{-1}, \texttt{edge}, \texttt{edge}^{-1}, \texttt{node}, \texttt{node}^{-1}\}$.

The nesting construction $[exp]$ is used to check for the existence of a path defined by expression exp. For instance, when evaluating nested expression $\texttt{next::}[exp]$ in a graph G, we retrieve the pair of nodes (x, y) such that there exists z with $(x, z, y) \in G$, and such that there is a path in G that follows expression exp starting in z. The formal semantics of nested regular path expressions is shown in Table 3. The semantics for the navigation axes of the form 'axis' and 'axis::a', as well as the concatenation, disjunction, and star closure of expressions, is defined as for the case of regular path expressions (see Table 2).

Table 3. Formal semantics of nested regular path expressions

$[\![\texttt{self}::[exp]]\!]_G = \{(x,x) \mid x \in \text{voc}(G) \text{ and there exists } z \text{ s.t. } (x,z) \in [\![exp]\!]_G\}$

$[\![\texttt{next}::[exp]]\!]_G = \{(x,y) \mid \text{there exist } z,w \text{ s.t. } (x,z,y) \in G \text{ and } (z,w) \in [\![exp]\!]_G\}$

$[\![\texttt{edge}::[exp]]\!]_G = \{(x,y) \mid \text{there exist } z,w \text{ s.t. } (x,y,z) \in G \text{ and } (z,w) \in [\![exp]\!]_G\}$

$[\![\texttt{node}::[exp]]\!]_G = \{(x,y) \mid \text{there exist } z,w \text{ s.t. } (z,x,y) \in G \text{ and } (z,w) \in [\![exp]\!]_G\}$

$[\![\text{axis}^{-1}::[exp]]\!]_G = \{(x,y) \mid (y,x) \in [\![\text{axis}::[exp]]\!]_G\}$ with axis $\in \{\texttt{next}, \texttt{node}, \texttt{edge}\}$

Example 6. Consider an RDF graph G storing information about transportation services between cities. As in Example 4, a triple (C_1, tc, C_2) in the graph indicates that there is a direct way of traveling from C_1 to C_2 by using the transportation company tc. Then the nested expression:

$$(\texttt{next}::\text{KoreanAir})^+/\texttt{self}::[(\texttt{next}::\text{AirFrance})^*/\texttt{self}::\text{Paris}]/(\texttt{next}::\text{KoreanAir})^+,$$

defines the pairs of cities (C_1, C_2) such that, there is a way of flying from C_1 to C_2 with KoreanAir with a stop in a city C_3 from which one can fly to Paris with AirFrance. Notice that $\texttt{self}::[(\texttt{next}::\text{AirFrance})^*/\texttt{self}::\text{Paris}]$ is used to test for the existence of a flight (that can have some stops) from C_3 to Paris with AirFrance.   □

Recall that rSPARQL was defined as the extension of SPARQL with regular path expressions in the predicate position of triple patterns. Similarly, *nested* SPARQL (or just nSPARQL) is defined as the extension of SPARQL with nested regular expressions in the predicate position of triple patterns. The following example shows the benefits of using nSPARQL when trying to obtain the RDFS evaluation of a pattern by directly traversing an RDF graph.

Example 7. Consider the SPARQL pattern $P = (?X, \text{works\_in}, ?Y)$. We have seen that it is not possible to obtain the RDFS evaluation of P with an rSPARQL pattern. Consider now the nested regular expression:

$$\texttt{next}::[(\texttt{next}::\text{sp})^*/\texttt{self}::\text{works\_in}]. \tag{3}$$

It defines the pairs (a, b) of resources in an RDF graph G such that, there exist a triple (a, x, b) and a path from x to works_in in G where every edge has label sp. The expression $(\texttt{next}::\text{sp})^*/\texttt{self}::\text{works\_in}$ is used to simulate the inference process in RDFS; it retrieves all the nodes that are *sub-properties* of works_in. Thus, expression (3) is exactly what we need to obtain the RDFS evaluation of pattern P. In fact, if G is the RDF graph in Figure 1 and Q the nSPARQL pattern:

$$Q = (?X, \texttt{next}::[(\texttt{next}::\text{sp})^*/\texttt{self}::\text{works\_in}], ?Y),$$

then we obtain

$$[\![Q]\!]_G = \begin{array}{|c|c|} \hline ?X & ?Y \\ \hline \text{Ronaldinho} & \text{Barcelona} \\ \hline \text{Sorace} & \text{Everton} \\ \hline \end{array}$$

This is exactly the RDFS evaluation of P in G, that is, $[\![P]\!]_G^{\text{rdfs}} = [\![Q]\!]_G$.   □

It turns out that nested expressions are the necessary ingredient to obtain the RDFS evaluation of SPARQL patterns by navigating RDF graphs. To show that this holds, consider the following *translation* function from elements in U to nested expressions:

$$
\begin{aligned}
\text{trans}(\mathsf{sc}) \quad &= (\mathsf{next::sc})^+ \\
\text{trans}(\mathsf{sp}) \quad &= (\mathsf{next::sp})^+ \\
\text{trans}(\mathsf{dom}) \quad &= \mathsf{next::dom} \\
\text{trans}(\mathsf{range}) &= \mathsf{next::range} \\
\text{trans}(\mathsf{type}) \quad &= (\ \mathsf{next::type}/(\mathsf{next::sc})^* \mid \\
&\qquad \mathsf{edge}/(\mathsf{next::sp})^*/\mathsf{next::dom}/(\mathsf{next::sc})^* \mid \\
&\qquad \mathsf{node}^{-1}/(\mathsf{next::sp})^*/\mathsf{next::range}/(\mathsf{next::sc})^*\ ) \\
\text{trans}(p) \quad &= \mathsf{next::}[(\mathsf{next::sp})^*/\mathsf{self::}p] \quad \text{for } p \notin \{\mathsf{sc}, \mathsf{sp}, \mathsf{range}, \mathsf{dom}, \mathsf{type}\}.
\end{aligned}
$$

By using the results of [18], it can be shown that for every SPARQL triple pattern of the form (x, a, y), where $x, y \in U \cup V$ and $a \in U$, it holds that:

$$
[\![(x, a, y)]\!]^{\mathrm{rdfs}}_G = [\![(x, \text{trans}(a), y)]\!]_G
$$

for every RDF graph G. That is, given an RDF graph G and a triple pattern t not containing a variable in the predicate position, it is possible to obtain the RDFS evaluation of t over G by navigating G through a nested regular expression (and without explicitly computing the closure of G).

Given that the syntax and semantics of SPARQL patterns are defined from triple patterns, the previous property also holds for SPARQL patterns including operators AND, OPT, UNION and FILTER. That is, if P is a SPARQL pattern constructed by using triple patterns from the set $(U \cup V) \times U \times (U \cup V)$, then there is an nSPARQL pattern Q such that for every RDF graph G, it holds that $[\![P]\!]^{\mathrm{rdfs}}_G = [\![Q]\!]_G$.

It should be noticed that, if variables are allowed in the predicate position of triple patterns, in general there is no hope to obtain the RDFS evaluation without computing the closure, since a triple pattern like $(?X, ?Y, ?Z)$ can be used to retrieve the entire closure of an RDF graph.

3.4   The Extra Expressive Power of Nested Regular Expressions

Nested regular expressions were designed to be expressive enough to capture the semantics of RDFS. Beside this feature, nested regular expressions also provide some other interesting features that give extra expressiveness to the language. With nested regular expressions, one is allowed to define complex paths by using concatenation, disjunction and star closure, over nested expressions. It is also allowed to use various levels of nesting in expressions. Note that these features are not needed in the translations presented in the previous section.

The following example shows that the extra expressiveness of nested regular expressions can be used to formulate interesting and natural queries, which cannot be expressed by using regular path expressions.

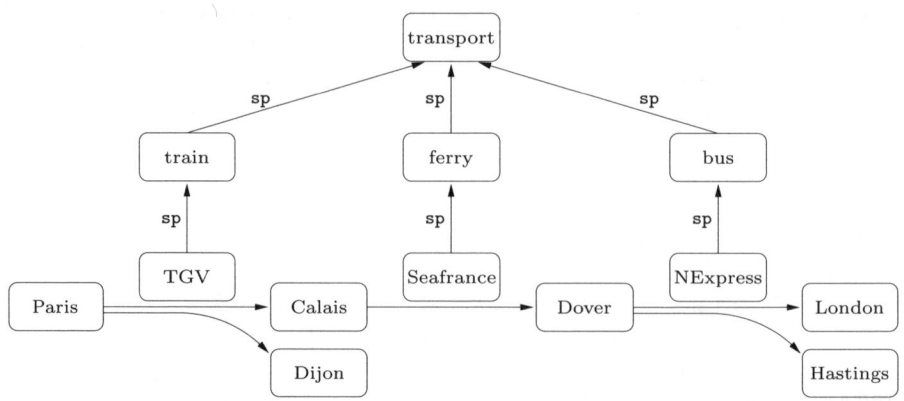

Fig. 4. An RDF graph storing information about transportation services between cities

Example 8. Consider the RDF graph with transportation information in Figure 4. As in the previous examples, if C_1 and C_2 are cities and (C_1, tc, C_2) is a triple in the graph, then there is a direct way of traveling from C_1 to C_2 by using the transportation company tc. For instance, (Paris, TGV, Calais) indicates that TGV provides a transportation service from Paris to Calais. In the figure, we also have extra information about the travel services. For example, TGV is a sub-property of train and then, if (Paris, TGV, Calais) is in the graph, we can infer that there is a train going from Paris to Calais.

If we want to know whether there is a way to travel from one city to another (without taking into consideration the kind of transportation), we can use the following expression:

$$(\mathtt{next}::[(\mathtt{next}::\mathtt{sp})^*/\mathtt{self}::\mathrm{transport}])^+.$$

Assume now that we want to obtain the pairs (C_1, C_2) of cities such that there is a way to travel from C_1 to C_2 with a stop in a city which is either London or is connected by a bus service with London. First, notice that the following nested expression checks whether there is a way to travel from C_1 to C_2 with a stop in London:

$$(\mathtt{next}::[(\mathtt{next}::\mathtt{sp})^*/\mathtt{self}::\mathrm{transport}])^+/\mathtt{self}::\mathrm{London}/$$
$$(\mathtt{next}::[(\mathtt{next}::\mathtt{sp})^*/\mathtt{self}::\mathrm{transport}])^+. \quad (4)$$

Thus, to obtain an expression for our initial query, we only need to replace $\mathtt{self}::\mathrm{London}$ in (4) by an expression that checks whether a city is either London or is connected by a bus service with London. The following expression can be used to test the latter condition:

$$(\mathtt{next}::[(\mathtt{next}::\mathtt{sp})^*/\mathtt{self}::\mathrm{bus}])^*/\mathtt{self}::\mathrm{London}. \quad (5)$$

Hence, by replacing self::London by (5) in nested regular expression (4), we obtain a nested regular expression for our initial query:

$$(\text{next}::[(\text{next}::\text{sp})^*/\text{self}::\text{transport}])^+/$$
$$\text{self}::[(\text{next}::[(\text{next}::\text{sp})^*/\text{self}::\text{bus}])^*/\text{self}::\text{London}] /$$
$$(\text{next}::[(\text{next}::\text{sp})^*/\text{self}::\text{transport}])^+. \quad (6)$$

Notice that the level of nesting of (6) is 2. If we evaluate (6) over the RDF graph in Figure 4, we obtain the pair (Calais, Hastings) as a possible answer since there is a way to travel from Calais to Hastings with a stop in Dover, from which there is a bus service to London.   □

4   Concluding Remarks

The problem of answering queries over RDFS is challenging, due to the existence of a vocabulary with a predefined semantics. Current approaches for this problem pre-compute the closure of RDF graphs. From a practical point of view, these approaches have several drawbacks, among others that they are not goal-oriented: although a query may need to scan a small part of the data, all the data is considered when computing the closure of an RDF graph.

In this paper, we propose an alternative approach to the problem of answering RDFS queries. We present a navigational language constructed from *nested regular expressions*, that can be used to obtain the answer to RDFS queries by navigating the input graph (without pre-computing the closure). Besides capturing the semantics of RDFS, nested regular expressions also provide some other interesting features that give extra expressiveness to the language. We think these features deserve further and deeper investigation.

Acknowledgments. The authors were supported by: Arenas – FONDECYT grant 1070732; Gutierrez – FONDECYT grant 1070348; Pérez – CONICYT Ph.D. Scholarship; Arenas, Gutierrez and Pérez – grant P04-067-F from the Millennium Nucleus Center for Web Research.

References

1. Alkhateeb, F., Baget, J., Euzenat, J.: Complex path queries for RDF. Poster paper in ISWC 2005 (2005)
2. Alkhateeb, F., Baget, J., Euzenat, J.: RDF with regular expressions. Research Report 6191, INRIA (2007)
3. Angles, R., Gutierrez, C.: Survey of graph database models. ACM Comput. Surv. 40(1), 1–39 (2008)
4. Anyanwu, K., Maduko, A., Sheth, A.: SemRank: ranking complex relationship search results on the semantic web. In: WWW 2005, pp. 117–127 (2005)
5. Anyanwu, K., Maduko, A., Sheth, A.: SPARQ2L: Towards Support for Subgraph Extraction Queries in RDF Databases. In: WWW 2007, pp. 797–806 (2007)

6. Brickley, D., Guha, R.V.: RDF Vocabulary Description Language 1.0: RDF Schema. W3C Recommendation (Feburary 2004), http://www.w3.org/TR/rdf-schema/
7. Broekstra, J., Kampman, A., van Harmelen, F.: Sesame: A generic architecture for storing and querying rdf and rdf schema. In: Horrocks, I., Hendler, J. (eds.) ISWC 2002. LNCS, vol. 2342, pp. 54–68. Springer, Heidelberg (2002)
8. Clark, J., DeRose, S.: XML Path Language (XPath). W3C Recommendation (November 1999), http://www.w3.org/TR/xpath
9. Furche, T., Linse, B., Bry, F., Plexousakis, D., Gottlob, G.: RDF Querying: Language Constructs and Evaluation Methods Compared. In: Barahona, P., Bry, F., Franconi, E., Henze, N., Sattler, U. (eds.) Reasoning Web 2006. LNCS, vol. 4126, pp. 1–52. Springer, Heidelberg (2006)
10. Gutierrez, C., Hurtado, C., Mendelzon, A.: Foundations of Semantic Web Databases. In: PODS 2004 (2004)
11. Harris, S., Gibbins, N.: 3store: Efficient bulk RDF storage. In: Proceedings of the 1st International Workshop on Practical and Scalable Semantic Systems (PSSS 2003), Sanibel Island, Florida, pp. 1–15 (2003)
12. Haase, P., Broekstra, J., Eberhart, A., Volz, R.: A Comparison of RDF Query Languages. In: McIlraith, S.A., Plexousakis, D., van Harmelen, F. (eds.) ISWC 2004. LNCS, vol. 3298, pp. 502–517. Springer, Heidelberg (2004)
13. Hayes, J., Gutierrez, C.: Bipartite Graphs as Intermediate Model for RDF. In: McIlraith, S.A., Plexousakis, D., van Harmelen, F. (eds.) ISWC 2004. LNCS, vol. 3298, pp. 47–61. Springer, Heidelberg (2004)
14. Hayes, P.: RDF Semantics. W3C Recommendation (February 2004), http://www.w3.org/TR/rdf-mt/
15. Kochut, K., Janik, M.: SPARQLeR: Extended Sparql for Semantic Association Discovery. In: Franconi, E., Kifer, M., May, W. (eds.) ESWC 2007. LNCS, vol. 4519, pp. 145–159. Springer, Heidelberg (2007)
16. Manola, F., Miller, E., McBride, B.: RDF Primer, W3C Recommendation (February 10, 2004), http://www.w3.org/TR/REC-rdf-syntax/
17. Mendelzon, A., Wood, P.: Finding Regular Simple Paths in Graph Databases. SIAM J. Comput. 24(6), 1235–1258 (1995)
18. Muñoz, S., Pérez, J., Gutierrez, C.: Minimal Deductive Systems for RDF. In: Franconi, E., Kifer, M., May, W. (eds.) ESWC 2007. LNCS, vol. 4519, pp. 53–67. Springer, Heidelberg (2007)
19. Olson, M., Ogbuji, U.: The Versa Specification, http://uche.ogbuji.net/tech/rdf/versa/etc/versa-1.0.xml
20. Pérez, J., Arenas, M., Gutierrez, C.: Semantics and Complexity of SPARQL. In: Cruz, I., Decker, S., Allemang, D., Preist, C., Schwabe, D., Mika, P., Uschold, M., Aroyo, L.M. (eds.) ISWC 2006. LNCS, vol. 4273, pp. 30–43. Springer, Heidelberg (2006)
21. Prud'hommeaux, E., Seaborne, A.: SPARQL Query Language for RDF. W3C Working Draft (March 2007), http://www.w3.org/TR/rdf-sparql-query/
22. Souzis, A.: RxPath Specification Proposal, http://rx4rdf.liminalzone.org/RxPathSpec

On RDF/S Ontology Evolution

George Konstantinidis[1,2], Giorgos Flouris[1],
Grigoris Antoniou[1,2], and Vassilis Christophides[1,2]

[1] Institute of Computer Science, FORTH, Greece
[2] Computer Science Department, University of Crete, Greece
{gconstan,fgeo,antoniou,christop}@ics.forth.gr

Abstract. The algorithms dealing with the incorporation of new knowledge in an ontology (ontology evolution) often share a rather standard process of dealing with changes. This process consists of the specification of the language, the determination of the allowed update operations, the identification of the invalidities that could be caused by each such operation, the determination of the various alternatives to deal with each such invalidity, and, finally, some selection mechanism for singling out the "best" of these alternatives. Unfortunately, most ontology evolution algorithms implement these steps using a case-based, ad-hoc methodology, which is cumbersome and error-prone. The first goal of this paper is to present, justify and make explicit the five steps of the process. The second goal is to propose a general framework for ontology change management that captures this process, in effect generalizing the methodology employed by existing tools. The introduction of this framework allows us to devise a whole class of ontology evolution algorithms, which, due to their formal underpinnings, avoid many of the problems exhibited by ad-hoc frameworks. We exploit this framework by implementing a specific ontology evolution algorithm for RDF ontologies as part of the FORTH-ICS Semantic Web Knowledge Middleware (SWKM).

1   Introduction

Change management is a key component of any knowledge-intensive application. The same is true for the Semantic Web, where knowledge is usually expressed in terms of ontologies and refined through various methodologies using *ontology evolution* techniques. The most critical part of an ontology evolution algorithm is the determination of *what* can be changed and *how* each change should be implemented. The main argument of this paper is that this determination can be split into the following 5 steps, which, although not explicitly stated, are shared by many ontology evolution tools:

1. *Model Selection.* The allowed changes, as well as the various alternatives for implementing each change, are constrained by the expressive power of the ontology representation model. Thus, the selection of the model may have profound effects on what can be changed, and how, so it constitutes an important parameter of the evolution algorithm.

V. Christophides et al. (Eds.): SWDB-ODBIS 2007, LNCS 5005, pp. 21–42, 2008.
© Springer-Verlag Berlin Heidelberg 2008

2. *Supported Operations.* In step 2, the supported change operations upon the ontology are specified.

3. *Validity Model.* Problems related to the validity of the resulting ontology may arise whenever a change operation is executed; such problems depend on the validity model assumed for ontologies.

4. *Invalidity Resolution.* This step determines, for each supported operation and possible invalidity problem, the different (alternative) actions that can be performed to restore the validity of the ontology.

5. *Action Selection.* During this step, a selection process is used to determine the most preferable among the various potential actions (that were identified in the previous step) for execution.

Unfortunately, most of the existing frameworks (e.g., [1,6,10,17]) address such ontology evolution issues in an ad-hoc way. As we will see in Section 3, this approach causes a number of problems (e.g., reduced flexibility, limited evolution primitives, non-faithful behavior etc), so evolution algorithms could benefit a lot from the formalization of the aforementioned change management process. In Section 2, we define ontology evolution and give a general overview of the state of the art in the field; this allows us to motivate our work and place it in its correct context. In Section 3, we describe four typical ontology evolution systems, namely OilEd [1], KAON [6], Protégé [10] and OntoStudio (formerly OntoEdit [17]); we show how these systems fit on the aforementioned five-step process and criticize the ad-hoc methodology that they employ to face these steps.

Section 4 introduces the general formal framework that we employ in order to model the various steps of this process. Our framework allows us to deal with arbitrary change operations (rather than a predetermined set). In addition, it considers all the invalidity problems that could, potentially, be caused by each change, and all the possible ways to deal with them. Finally, it provides a parameterizable method to select the "best" out of the various alternative options to deal with an invalidity, according to some metric. The formal nature of the process allows us to avoid resorting to the tedious and error-prone manual case-based reasoning that is necessary in other frameworks for determining invalidities and solutions to them, and provides a uniform way to select the "best" option out of the list of available ones, using some total ordering. Our framework can be used for several different declarative ontological languages and semantics; however, for implementation and visualization purposes, we instantiate it for the case of RDF, under the semantics described in [13].

Finally, in Section 5, we exhibit the merits of our framework via the development of a general-purpose algorithm for ontology evolution. This algorithm has general applicability, but we demonstrate how it can be employed for the RDF case. Then, we specialize our approach for the case of RDF and devise a number of special-purpose algorithms for coping with RDF changes (similar to the existing ad-hoc ontology evolution algorithms), which sacrifice generality for efficiency; the main advantage of such special-purpose algorithms with respect to the standard ad-hoc methodologies is that, due to their formal underpinnings and their proven compatibility with the general framework, they enjoy the same interesting properties.

The above algorithms are currently being implemented as part of the FORTH-ICS Semantic Web Knowledge Middleware (SWKM), which provides generic services for acquiring, refining, developing, accessing and distributing community knowledge. The SWKM is composed of four services, namely the Comparison Service (which compares two RDF graphs, reporting their differences), the Versioning Service (which handles and stores different versions of RDF graphs), the Registry Service (which is used to manipulate metadata information related to RDF graphs) and the Change Impact Service (which deals with the evolution of RDF graphs). The SWKM is backed up by a number of more basic services (Knowledge Repository Services) which allow basic storage and access functionalities for RDF graphs[1]. This paper describes the algorithms we employ for the Change Impact Service of SWKM, as well as the underlying theoretical background of the service.

2   Related Work and Motivation

2.1   Short Literature Review

Ontology evolution deals with the incorporation of new knowledge in an ontology; more accurately, the term refers to *the process of modifying an ontology in response to a certain change in the domain or its conceptualization* [5]. Ontology evolution is an important problem, as the effectiveness of an ontology-based application heavily depends on the quality of the conceptualization of the domain by the underlying ontology, which is directly affected by the ability of an evolution algorithm to properly adapt the ontology both to changes in the domain (as ontologies often model dynamic environments) and to changes in the domain's conceptualization (as no conceptualization can ever be perfect) [5].

In order to tame the complexity of the problem, six phases of ontology evolution have been identified in [14], occurring in a cyclic loop. Initially, we have the *change capturing* phase, where the changes to be performed are determined; these changes are formally represented during the *change representation* phase. The third phase is the *semantics of change* phase, in which the effects of the change(s) to the ontology itself are determined; during this phase, possible problems that might be caused to the ontology by these changes are also identified and resolved. The *change implementation* phase follows, where the changes are physically applied to the ontology, the ontology engineer is informed of the changes and the performed changes are logged. These changes need to be propagated to dependent elements; this is the role of the *change propagation* phase. Finally, the *change validation* phase allows the ontology engineer to review the changes and possibly undo them, if desired. This phase may uncover further problems with the ontology, thus initiating new changes that need to be performed to improve

---

[1] For more details on the architecture of the SWKM, see:
http://athena.ics.forth.gr:9090/SWKM/index.html

the conceptualization; in this case, we need to start over by applying the change capturing phase of a new evolution process, closing the cyclic loop.

This paper focuses on the second and third phase (change representation and semantics of change), which are the most critical for ontology evolution [12]. Notice that during the change representation phase we determine the *requested* change (i.e., *what* should be changed), whereas during the semantics of change we determine the *actual* change (i.e., *how* the change should be performed). With respect to the five-step process described in Section 1, the change representation phase corresponds to the first two steps of our framework, whereas the semantics of change phase corresponds to the last three steps.

There is a rich literature that deals with the problem of ontology evolution. In general, two major research paths can be identified [5]. The first focuses on aiding the user performing changes in ontologies through some intuitive interface that provides a number of useful editing features; such tools resemble an ontology editor (and some of them are indeed ontology editors [15]), even though they often provide many more features than a simple ontology editor would. The second research path focuses on the development of automated methods to determine the effects and side-effects of any given update request (which correspond to phases 2 and 3 of [14]); this approach often borrows ideas from the related, and much more mature, discipline of belief change [7].

The first class of tools is more mature at the moment, but the second approach seems more interesting from a research point of view, as well as more promising; for this reason, it is gaining increasing attention during the last few years [5]. The two research paths are complementary, as results from the second could be applied to the first in order to further improve the quality of the front-end editing tools; similarly, automated approaches are of little use unless coupled with tools that address the practical issues related to evolution, like support for multi-user environments, transactional issues, change propagation, intuitive visual interfaces etc (i.e., the remaining four phases of [14]).

2.2 Motivation

Unfortunately, the above complementarity is not sufficiently exploited. Automated approaches (second research path) seem, in general, detached from real problems and are not easily adaptable for use in an ontology evolution tool; to our knowledge, there is no implemented tool that uses one of the algorithms developed by such approaches. On the other hand, editor-like tools (first research path) do not provide enough automation and employ ad-hoc methodologies to deal with the problems raised during an update operation; such ad-hoc methodologies cause several problems that are thoroughly discussed in Section 3.

Our approach is motivated by the need to develop a formal framework that will lead to an easily implementable ontology evolution algorithm. We would like our approach to enjoy the formality of the second class of tools, and use this formality as a basis that will provide guarantees related to the behavior of the implemented system, thus avoiding the problems related to the ad-hoc nature of existing practical methodologies. This paper is an attempt towards this end. In this respect, the work presented here lies somewhere between the two research paradigms described above, sharing properties with both worlds.

More specifically, our approach could be viewed as belonging to the second class of works, in the sense that it results to a formal, theoretical model to address changes. This model is based on a formal framework that is used to describe the process of ontology evolution as addressed by current editor-like tools (so it is also related to the first class of works), and allows us to develop an abstract, general-purpose algorithm that provably performs changes in an automated and rational way for a variety of languages, under different parameters (validity model and ordering). Like other works of the second research path above, our work is focused on the "core" of the ontology evolution problem, namely the change representation and semantics of change phases. Issues related to change capturing, implementation of changes, transactional issues, change propagation, visualization, interfaces, validation of the resulting ontology etc are not considered in this paper.

On the other hand, our approach could be viewed as belonging to the first class of tools, in the sense that it results to an implemented tool, namely the Change Impact Service of the SWKM. Our general-purpose algorithm can be applied for any particular language and set of parameters that is useful for practical purposes; for the purposes of SWKM we set these parameters so as to correspond to the RDF language under the semantics described in [13]. Fixing these parameters also allows us to better present our approach, as well as to evaluate and verify its usefulness towards the aim of implementing an ontology evolution tool. In addition to the implementation of the general-purpose algorithm, our formal framework allows the development (and implementation) of special-purpose algorithms which are more suited for practical purposes; such algorithms provably exhibit the same behavior as the general-purpose one, so we can have formal guarantees as to their expected output. For reasons explained in Section 5, both the general-purpose and the special-purpose algorithms are implemented for the Change Impact Service of SWKM.

3 Evolution Process in Current Systems

In this section, we elaborate on the five steps we described in Section 1 and describe how some typical ontology evolution tools ([1,6,10,17]) fit into this five-step process. In addition, we point out the problems that the ad-hoc implementation of these tools causes, and show how such problems could be overcome through the use of a formal framework, like the one described in Section 4.

3.1 Model Selection and Supported Operations

Obviously, the first step towards developing an evolution algorithm is the determination of the underlying representation model for the evolving ontology; this is what we capture in the first step of our 5-step process. Most systems assume a language supporting the basic constructs used in ontology development, like class and property subsumption relationships, instantiation relationships and domain and range restrictions for properties.

The selection of the representation model obviously affects (among other things) the operations that can be supported; for example, OntoStudio [17] does not support property subsumption relations so all related changes are similarly overruled. Further restrictions to the allowable changes may be introduced by various design decisions, which may disallow certain operations despite the fact that they could, potentially, be supported by the underlying ontology model. For example, OntoStudio does not allow the manipulation of implicit knowledge, whereas OilED [1] does not support any operation that would render the ontology invalid (i.e., it does not take any actions to restore validity, but rejects the entire operation instead). The determination of the allowed (supported) update operations constitutes the second step of our 5-step process.

According to [14,15], change operations can be classified into elementary (involving a change in a single ontology construct) and composite ones (involving changes in multiple constructs), also called atomic and complex in [16]. Elementary changes represent simple, fine-grained changes; composite changes represent more coarse-grained changes and can be replaced by a series of elementary changes. Even though possible, it is not generally appropriate to use a series of elementary changes to replace a composite one, as this might cause undesirable side-effects [14]; the proper level of granularity should be identified in each case. Examples of elementary changes are the addition and deletion of elements (concepts, properties etc) from the ontology. There is no general consensus in the literature on the type and number of composite changes that are necessary. In [14], 12 different composite changes are identified; in [11], 22 such operations are listed; in [16] however, the authors mention that they have identified 120 different interesting composite operations and that the list is still growing! In fact, since composite operations can involve changes in an arbitrary number of constructs, there is an infinite number of them. Although composite operations can, in general, be decomposed into a series of elementary ones, for ad-hoc systems this is not of much help, as the decomposition of a non-supported operation into a series of supported ones (even if possible) should be done manually.

The above observations indicate an important inherent problem with ad-hoc algorithms, namely that they can only deal with a predefined (and finite) set of supported operations, determined at design time. Therefore, any such algorithm is limited, because it can only support some of the potential changes upon an ontology, namely the ones that are considered more useful (at design time) for practical purposes, and, thus, supported.

3.2 Validity Model and Invalidity Resolution

It is obvious that a user expects his update request to be executed upon the ontology. Thus, it is necessary for the resulting ontology to actually implement the change operation originally requested, i.e., that the actual changes performed upon the ontology are a superset of the requested ones; this requirement will be called *success*.

The naive way to implement an update request upon an ontology would be to simply execute the request in a set-theoretic way. That would guarantee the satisfaction of the above principle (success); nevertheless, this would not be acceptable in most cases, because the resulting ontology could be invalid in some sense (e.g., if a class is removed, it does not make sense to retain subsumption relationships involving that class). Thus, another basic requirement for a change operation is that the result of its application should be a *valid* ontology, according to some validity model. This requirement is necessary in order for the resulting ontology to make sense.

Both the above principles are inspired by research on the related field of belief revision [3,7], in which they are known as the Principle of Validity and Principle of Success respectively. The Principle of Success is well-defined, in the sense that we can always verify whether it is satisfied or not. The Principle of Validity however, depends on some underlying validity model, which is not necessarily the same for all languages (ontology models) and/or ontology evolution systems. Thus, each system should define the validity model that it uses. For example, do we accept cycles in the IsA hierarchy? Do we allow properties without a range/domain, or with multiple ranges/domains? Such decisions are included in the validity model determined in step 3 of our 5-step process. Notice that the validity model has a different purpose than the ontology model: the ontology model is used to determine what constructs are available for use in an ontology (e.g., IsAs), whereas the validity model determines the valid combinations of constructs in an ontology (e.g., by disallowing cyclic IsAs).

Determining how to satisfy the Principles of Success and Validity during a change operation is not trivial. The standard process in this respect is to execute the original update request in a naive way (i.e., by executing plain set-theoretic additions and deletions), followed by the initiation of additional change operations (called *side-effects*) that would guarantee validity. In principle, there is no unique set of side-effects that could be used for this purpose: in some cases, there is more than one alternatives, whereas in others there is none. The latter type of updates (i.e., updates for which it is not possible for both Success and Validity to be satisfied) are called *infeasible* and should be rejected altogether. For example, the request to remove a class, say C, and add a subsumption relationship between C and D at the same time would be infeasible, because executing both operations of the composite update would lead the ontology to an invalid state (because a removed class C cannot be subsumed by another class) and it can be easily shown that there is no way (i.e., side-effects) to restore validity without violating success for this update. The determination of whether an update is infeasible or not, as well as of the various alternative options (for side-effects) that we have for guaranteeing success and validity (for feasible updates) constitutes the fourth step of our 5-step process.

Let us consider the change operation depicted in Figure 1(a), where the ontology engineer expresses the desire to delete a class (B) which happens to subsume another class (C). It is obvious that, once class B is deleted, the IsAs relating B with A and C would refer to a non-existent class (B), so they should be removed;

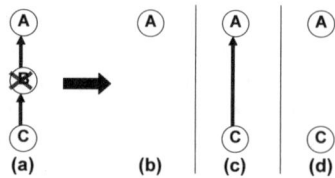

Fig. 1. Three alternatives for deleting a class

the validity model should capture this case, and attempt to resolve it. One possible result of this process, employed by Protégé [10], is shown in Figure 1(b); in that evolution context, a class deletion causes the deletion of its subclasses as well. This is not the only possibility though; Figures 1 (c) and (d), present other potential results of this operation, where in (c), B's subclasses are re-connected to its father, while in (d), the implicit IsA from C to A is not taken into account. KAON [6], for example, would give either of the three as a result, depending on a user-selected parameter.

In this particular example, both KAON and Protégé detect the invalidity caused by the operation and actively take action against it; however, the validity model employed by different systems may be different in general. Moreover, notice that an invalidity is not caused by the operation itself, but by the combination of the current ontology state and the operation (e.g., if B was not in any way connected to A and C, its deletion would cause no problems). Therefore, in order for a mechanism to propose solutions against invalidities, both the ontology and the update should be taken into account. Notice that the mechanism employed by Protégé, in Figure 1, identifies only a single set of side-effects, while KAON identifies three different reactions. This is not a peculiarity of this example; the invalidity resolution mechanism employed by Protégé identifies only a single solution per invalidity; this is not true for KAON and OntoStudio.

3.3 Action Selection

Since, in the general case, there are several alternative ways (i.e., sets of side-effects) to guarantee success and validity, we need a mechanism that would allow us to select one of the alternatives for implementation (execution). This constitutes the last component of an evolution algorithm (step 5). Such a mechanism is "pre-built" into systems that identify only a single possible action, like Protégé, but can be also parameterizable. KAON, for example, provides a set of options (called *evolution strategies*) which allow the ontology engineer to tune the system's behavior and, implicitly, indicate what is the appropriate invalidity resolution action for implementation per case. OntoStudio provides a similar customization over its change strategies.

Notice that our preference for the result of an operation reflects in a preference among the possible side-effects of the operation. For instance, if we prefer the result of Figure 1 (c), we can equivalently say that we prefer the (explicit) addition of the (implicit) subsumption relation shown in (c) together with the deletion of the two initial IsAs as a side-effect to this operation, over the deletion of the two initial IsAs and class C, shown in (b), or just the deletion of the two IsAs, as in (d). Therefore, the evolution process can be tuned by introducing a preference ordering upon the operations' side-effects that would dictate the related choice (evolution strategy). Given that the determination of the alternative side-effects depends on both the update and the ontology, there is an infinite number of different potential side-effects that may have to be compared. Thus, we are faced with the challenge of introducing a preference mechanism that will be able to compare any imaginable pair of side-effects.

It is worth noting here the connection of this preference ordering with the well-known belief revision Principle of Minimal change [3] which states that the resulting ontology should be as "close" as possible to the original one. In this sense, the preference ordering could be viewed as implying some notion of relative distance between different results and the original ontology, as identified by the preference between these results' corresponding side-effects.

3.4 Discussion

To the best of authors' knowledge, all currently implemented systems employ ad-hoc mechanisms to resolve the issues described above. The designers of these systems have determined, in advance (i.e., at design time), the supported operations, the possible invalidities that could occur per operation, the various alternatives for handling any such possible invalidity, and have already pre-selected the preferable option (or options, for flexible systems like KAON) for implementation per case; this selection (or selections) is hard-coded into the systems' implementations.

This approach causes a number of problems. First of all, each invalidity, as well as each of the possible solutions to each one, needs to be considered individually, using a highly tedious, manual case-based reasoning which is error-prone and gives no formal guarantee that the cases and options considered are exhaustive. Similarly, the nature of the selection mechanisms cannot guarantee that the selections (regarding the proper side-effects) that are made for different operations exhibit a faithful overall behavior. This is necessary in the sense that the side-effect selections made in different operations (and on different ontologies) should be based on an operation-independent "global policy" regarding changes. Such a global policy is difficult to implement and enforce in an ad-hoc system.

Such systems face a lot of limitations due to the above problems. For example, OilED deals only with a very small fraction of the operations that could be defined upon its modeling, as any change operation that would be triggering side-effects is unsupported (e.g., the operation of Figure 1 is rejected). In Protégé, the design choice to support a large number of operations has forced its designers to limit the flexibility of the system by offering only one way of realizing a change; in OntoStudio, they are relieved of dealing with (part of) the complexity of the aforementioned case-based reasoning as the severe limitations

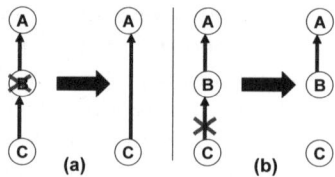

Fig. 2. Implicit knowledge handling in KAON

on the expressiveness of the underlying model constrain drastically the number of supported operations and cases to consider. Finally, in KAON, some possible side-effects are missing (ignored) for certain operations, while the selection process implied by KAON's parameterization may exhibit invalid or non-uniform behavior in some cases. As an example, consider Figure 2, in which the same evolution strategy was set in both (a) and (b); despite that, the implicit IsA from C to A is only considered (and retained) in case (a).

Table 1 summarizes some of the key features of ontology evolution systems, categorized according to the 5-step process introduced in this paper, and shows how each step is realized in each of the four systems discussed here, as well as in the Change Impact Service of SWKM, described in Sections 4, 5 below.

We argue that many of the problems identified in this section could be resolved by introducing an adequate evolution framework that would allow the description of an algorithm in more formal terms, as a modular sequence of choices regarding the ontology model used, the supported operations, the validity model, the identification of plausible side-effects and the selection mechanism. Such a framework would allow justified reasoning on the system's behavior,

Table 1. Summary of ontology evolution tools

			Protégé	KAON	OntoStudio	OilED	SWKM
Change Representation	Fine-grained Model (Step 1)		✓	✓	×	✓	✓
	Supported Operations (Step 2)	Elementary	✓	✓	✓	×	✓
		Composite	×	×	×	×	✓
Semantics of Change	Validity Model (Step 3)	Faithful	×	×	✓	✓	✓
		Complete	×	×	✓	×	✓
	Invalidity Resolution (Step 4)	No alternatives				✓	
		One alternative	✓				
		Many alternatives		✓			
		All alternatives			✓		✓
	Selection Mechanism (Step 5)	None	✓			✓	
		Per-case		✓	✓		
		Globally					✓

without having to resort to a case-by-case study of the various possibilities. To the best of the authors' knowledge, there is no implemented system that follows this policy. In Section 4, we describe such a framework and specialize it for RDF ontologies.

4   A Formal Framework for RDF/S Ontology Evolution

Our evolution framework consists of a fine-grained modeling of ontologies (step 1), a description of how both elementary and composite operations can be handled in a uniform way (step 2), a validity model formalized using integrity rules (step 3), which also allow us to document how side-effects are generated (step 4), and, finally, a selection mechanism based on an ordering that captures the Principle of Minimal Change (step 5). This framework will be instantiated to refer to RDF updating, but can be used for many different declarative languages, by tuning the various parameters involved.

4.1   Model Selection, Supported Operations and Validity Model

The representation model we use in this paper is the RDF language, in particular the model described in [13]. For ease of representation, RDF constructs will not be represented in the standard way, but we will use an alternative representation, which, in short, amounts to mapping each statement of RDF to a First-Order Logic (FOL) predicate (see Table 2); this way, a class IsA between A and B, for example, would be mapped to the predicate: $C\_IsA(A, B)$, while a triple denoting that the domain of a property, say P, is C, would be denoted by $Domain(P, C)$. Note that the standard alternative mapping (e.g., for IsA: $\forall x A(x) \rightarrow B(x)$) does not allow us to map assertions of the form "C is a class", and, consequently, does not allow us to handle operations like the addition or removal of a class, property, or instance (see [4] for more details on this issue). Notice that the same representation pattern can be used for other declarative languages as well, even though it is more suitable for simpler ones [4].

Table 2. Representation of RDF facts using FOL predicates

RDF triple	Intuitive meaning	Predicate
C rdf:type rdfs:Class	C is a class	$CS(C)$
P rdf:type rdf:Property	P is a property	$PS(P)$
x rdf:type rdfs:Resource	x is a class instance	$CI(x)$
P rdfs:domain C	domain of property	$Domain(P, C)$
P rdfs:range C	range of property	$Range(P, C)$
C_1 rdfs:subClassOf C_2	IsA between classes	$C\_IsA(C_1, C_2)$
P_1 rdfs:subPropertyOf P_2	IsA between properties	$P\_IsA(C_1, C_2)$
x rdf:type C	class instantiation	$C\_Inst(x, C)$
x P y	property instantiation	$PI(x, y, P)$

We equip our FOL with closed semantics, i.e., admit the *closed world assumption* (CWA). This means that, for a set S and a formula p, if $S \nvdash p$, then $S \vdash \neg p$. We overload $\vdash$ relation so as to be applicable between two sets as well: for two sets S, S' it holds that $S \vdash S'$ iff $S \vdash p$ for all $p \in S'$. Let us denote by L the set of ground facts allowed in our model (e.g., $C\_IsA(A, B), \neg CS(C)$), and L^+ the set of positive ground facts of L (e.g., $C\_IsA(A, B)$).

An ontology is represented as a set of positive ground facts only, so an ontology is any set $O \subseteq L^+$. Given CWA, the definition of an ontology and FOL semantics, it follows that: (a) an ontology is always consistent (in the standard FOL sense), (b) a positive ground fact is implied by an ontology iff it is contained in it, and, (c) a negative ground fact is implied by an ontology iff its positive counterpart is not contained in it.

An *update* is any set of positive and/or negative ground facts, so an update is any set $U \subseteq L$. According to the Principle of Success, an update should be implemented upon the ontology. Implementing a positive ground fact contained in an update is easy: all we have to do is add it to the ontology. However, this is not true for negative ground facts, because negative ground facts cannot be contained in an ontology, by definition. By CWA and the property (c) above, we conclude that "including" a negative ground fact in an ontology is equivalent to removing its positive counterpart. Given this analysis, we conclude that positive ground facts in an update correspond to additions, while negative ones correspond to removals. This way of viewing updates allows us to express essentially any operation, because any operation can be expressed as a set of additions and/or removals of ground facts in our model. Thus, we put no constraints on the allowed (supported) update operations.

Our framework needs also to define its validity model in a formal way. Validity can in general be formalized using a set of integrity constraints (rules) upon the ontology; therefore, a validity model is a set R of generic FOL formulas, which correspond to the axiomatization of the constraints of the model. For technical reasons that will be made apparent later, we constrain R to contain only "$\forall \exists$" formulas. Notice that the validity constraints should: (a) capture the notion of validity in the standard sense (e.g., that class subsumptions should be applied between classes in the ontology) and (b) encode the semantics of the various constructs of the underlying language (RDF in our case), which are not carried over during the transition to FOL (e.g., IsA transitivity) [4]. The latter type of constraints is very important, in the sense that it forces an ontology to contain all its implicit knowledge as well in order to be valid.

Similar to our approach, the authors of [8] consider the case of updating a set of facts representing a knowledge base, under a set of well-formed constraints on this base. However this work supports rather naïve changes as it does not consider any side-effects for a change (storing the updates that violate any rules as *exceptions* to the latter) nor composite updates. So, instead of implementing a more sophisticated change mechanism the authors of [8] emphasize on minimizing the size of the knowledge base, in the face of an update. Another work which considers updating structured data under constraints is presented in [2], where

Table 3. Indicative list of validity rules

Rule ID/Name	Integrity Constraint	Intuitive Meaning
R3 *Domain* Applicability	$\forall x, y : Domain(x, y) \rightarrow PS(x) \wedge CS(y)$	Domain applies to properties; the domain of a property is a class
R5 *C_IsA* Applicability	$\forall x, y : C\_IsA(x, y) \rightarrow CS(x) \wedge CS(y)$	Class IsA applies between classes
R12 *C_IsA* Transitivity	$\forall x, y, z : C\_IsA(x, y) \wedge C\_IsA(y, z) \rightarrow C\_IsA(x, z)$	Class IsA is Transitive

XML documents are automatically evaluated against a set of rules they should adhere to. However in case of invalidities, the process of updating the documents accordingly is left to be done manually. Therefore both of these works are essentially different from our approach as we develop an automated, parameterizable to its change policy, change mechanism, under a certain validity context (set of rules).

Table 3 contains an indicative list of the rules we use for RDF [13] (see also [9] for a similar effort). Notice that the rules presented are only a parameter of the model; our framework does not assume any particular set of rules (in the same sense that it does not assume any particular ontology representation language). However, the task of defining the respective rules becomes increasingly complex as the expressive power of the underlying logic increases, so this technique is more useful for less expressive languages (like RDF) [4].

4.2 Formalizing Our Model

We now have all the necessary ingredients for our formal definitions. Initially, an update algorithm can be formalized as a function mapping an ontology (i.e., a set of positive ground facts) and an update (i.e., a set of positive and negative ground facts) to another ontology. Thus:

Definition 1. *An update algorithm is a function* $\bullet : L^+ \times L \mapsto L^+$.

An ontology is *valid* iff it satisfies the rules of the validity model R, i.e., iff it implies all rules in R. Thus:

Definition 2. *An ontology O is valid, per the validity model R, iff $O \vdash R$.*

As already mentioned, the Principle of Success implies that all positive ground facts in an update should be included in the result, whereas the positive counterparts of the negative ground facts in an update should not. Thus, any (positive or negative) ground fact p in an update U should be implied by the result of the change operation. Of course, this is true for feasible updates; for infeasible updates, by definition, there is no valid ontology that satisfies the above requirement. Therefore:

Definition 3. *An update U is called feasible, per the validity model R, iff there is a valid ontology O (O ⊢ R) such that O ⊢ U. An update U is called infeasible iff it is not feasible.*

Definition 4. *Consider a language L, a set of validity rules R and an update algorithm* • : $L^+ \times L \mapsto L^+$. *Then:*

- *The algorithm • satisfies the Principle of Success iff for all valid ontologies $O \subseteq L^+$ and all feasible updates $U \subseteq L$, it holds that $O \bullet U \vdash U$.*
- *The algorithm • satisfies the Principle of Validity iff for all valid ontologies $O \subseteq L^+$ and all feasible updates $U \subseteq L$, it holds that $O \bullet U$ is a valid ontology.*

Notice that the above definition does not handle the cases where the input ontology is not valid to begin with, or when the update is infeasible; these are limit cases that will be handled separately later.

4.3   Invalidity Resolution and Action Selection

As already mentioned, the raw application of an update would guarantee success but could often violate validity (i.e., it could violate an integrity constraint). For example, under the validity context of Table 3, the raw application of the class deletion of Figure 1 would violate rule R5. In such cases, we need to determine the various options that we have in order to resolve the invalidity.

The formalization of the validity model using rules has the important property that, apart from detecting invalidities, it also provides a straightforward methodology to determine the various available options for resolving them. In effect, the rules themselves and the FOL semantics indicate the appropriate side-effects to be taken when an invalidity is detected. In the example with the class deletion (Figure 1), rule R5 implies that, in order to restore validity after the removal of class B (denoted by $\neg CS(B)$), we must delete the IsAs involving B.

In the general case, detecting and restoring an invalidity would require a FOL reasoner; however, our assumption that an ontology is a set of positive ground facts and that a rule is a "∀∃" formula, allows us to develop a much more efficient way. In particular, a "∀∃" rule can be equivalently rewritten as the conjunction of a set of *subrules*, where each subrule is a formula of the form ∀⋁∃ (see Table 4). Thus, by definition, an ontology O is valid iff it implies all subrules of all rules of the validity model. A subrule is implied by O iff, for all possible variables, at least one of the constituents of the disjunction is true (i.e., implied). Thus, a subrule can be violated iff a previously true constituent of the subrule is, due to the update, rendered false (i.e., not implied) and there is no other true constituent of the subrule. Similarly, the possible ways to render a violated subrule true should be chosen among all the constituents of the subrule, i.e., we should select one of the constituents of the subrule to be rendered true (through a side-effect); notice that the selected constituent should not be the one that was rendered false by the update itself (or else we would violate success).

Let us explain this process using an example. Consider rule R5, which is broken down into two subrules, as shown in Table 4. Let's consider subrule R5.1;

Table 4. Breaking rules into subrules

Rule ID/Name	Subrules of the rule
R3 *Domain* Applicability	$R3.1 : \forall x, y : \neg Domain(x, y) \vee PS(x)$
	$R3.2 : \forall x, y : \neg Domain(x, y) \vee CS(y)$
R5 *C_IsA* Applicability	$R5.1 : \forall x, y : \neg C\_IsA(x, y) \vee CS(x)$
	$R5.2 : \forall x, y : \neg C\_IsA(x, y) \vee CS(y)$
R12 *C_IsA* Transitivity	$R12.1 : \forall x, y, z :$
	$\neg C\_IsA(x, y) \vee \neg C\_IsA(y, z) \vee C\_IsA(x, z)$

this subrule is satisfied iff for all variables x, y, it either holds that $\neg C\_IsA(x, y)$, or it holds that $CS(x)$. If we remove a class (say B, denoted by $\neg CS(B)$) which previously existed in the ontology (cf. Figure 1), we should verify that subrule R5.1 is still true. This practically amounts to verifying that no class IsA starting from B exists in the ontology, i.e., that $\neg C\_IsA(B, y)$ is true for all y. If any such y exists (say $y = C$), then we must remove the respective IsA (i.e., $\neg C\_IsA(B, C)$ should be recorded as a side-effect).

Rule R12 is similar: R12.1 (which is the only subrule of R12) can be violated by, e.g., the addition of an IsA (say $C\_IsA(C, B)$). This could happen if, for example, an ontology contains $C\_IsA(B, A)$, but not $C\_IsA(C, A)$ (cf. Figure 3). To see this, set $x = C, y = B, z = A$ in R12.1, Table 4. The difference with the previous case is that now the violation can be restored in two different ways: either by removing $C\_IsA(B, A)$, or by adding $C\_IsA(C, A)$ (i.e., either $\neg C\_IsA(B, A)$ or $C\_IsA(C, A)$ could be selected as side-effects).

Notice that the selected side-effects are updates themselves, so they are enforced upon the ontology by being executed along with the original update; moreover, they could, just like any update, cause additional side-effects of their own. Another important remark is that, in some cases (e.g., R5.1), the invalidity resolution mechanism gives a straightforward result, in the sense that we only have one option to break the invalidity; in other cases (e.g., R12.1), we may have more than one alternative options. In the cases where we have different alternative sets of side-effects to select among, a mechanism to determine the "best" option, according to some metric, should be in place. In Section 3, we showed that our "preference" among the side-effects can be encoded using an ordering; given such an ordering (say $<$), all we need to do is find the minimal set of side-effects (with respect to $<$) among all possible ones and implement it.

As usual, our framework does not depend on any particular ordering. For technical reasons however, not all orderings can be employed for this purpose. In particular, to guarantee the rationality of the results, the ordering should depend on the underlying ontology as well (e.g., it is generally accepted that the removal of a general class is more "severe" than the deletion of a more specific class, but this criterion implies knowing the position of the class in the class hierarchy of the ontology). In addition, the ordering should be transitive and total; furthermore, it should be monotonic with respect to $\subseteq$ (i.e., $U \subseteq U'$ implies $U \leq U'$). Moreover, it should not be affected by void changes: for example, the addition of class C is a void operation in an ontology that already contains C and the removal of class D is a void operation in an ontology that does not contain D, so the inclusion of $CS(C)$ (or $\neg CS(D)$, respectively) in the side-effects of an update upon the above ontologies should not affect the "mildness" of the update. Finally, the ordering should be antisymmetric, modulo void operations (i.e., two updates have the same "mildness" iff their non-void operations are identical). We will call *update generating* an ordering satisfying these properties.

In our implementation, the proposed ordering is based on the ordering shown in Table 5 among the 18 positive and negative predicates. This ordering is expanded to refer to updates (i.e., sets of ground facts) using the general idea that an update U_1 is "preferable" or "better" than U_2 (denoted by $U_1 < U_2$) iff the "worst" predicate used in U_1, is "better" than the "worst" predicate used in U_2 where the predicates' relative preference is determined by the order shown in Table 5. Ties are resolved using cardinality considerations and/or the relative "importance" of the predicate's arguments in the original ontology, according to certain rules that determine "importance". Further details are omitted due to space limitations. It can be proven that our ordering is update-generating.

Table 5. Ordering of predicates

$PI < C\_Inst < P\_IsA < C\_IsA < \neg PI < \neg C\_Inst < \neg P\_IsA < \neg C\_IsA < \neg Domain < \neg Range < \neg CI < \neg PS < \neg CS < Domain < Range < CI < PS < CS$

4.4 Rational Ontology Evolution Algorithms

Now consider an update U applied upon an ontology O per the update algorithm $\bullet$, returning $O \bullet U$. The question is, what were the effects and side-effects that were applied upon O to get $O \bullet U$? The restriction that ontologies contain only positive ground facts is extremely helpful in this respect too. In particular, we can define the *Delta* between two ontologies as follows:

Definition 5. *Consider two ontologies $O_1, O_2 \subseteq L^+$. The Delta between O_1 and O_2 is defined as $Delta(O_1, O_2) = \{p \mid p \in O_2 \setminus O_1\} \cup \{\neg p \mid p \in O_1 \setminus O_2\}$.*

Notice that the result of *Delta* is an update, i.e., $Delta(O_1, O_2) \subseteq L$; given the above definition, the actual set of effects and side-effects that were applied upon O to get $O \bullet U$ is just $Delta(O, O \bullet U)$. Notice that $Delta(O, O \bullet U)$ will just return the non-void operations that led from O to $O \bullet U$; this is not a problem, as void operations do not affect the ordering. Given this *Delta* function, the Principle of Minimal Change can be formalized by requiring that an update algorithm should return an ontology $O \bullet U$ such that $Delta(O, O \bullet U)$ is minimal compared to $Delta(O, O')$ for all other possible results O'.

Of course, we need to specify what are the other "possible results", which, as already mentioned, are the ones that satisfy the Principles of Success and

Validity. Thus, our formal definition of the Principle of Minimal Change should be coupled with the other principles. We will therefore define a *rational* update algorithm to be one that satisfies the Principles of Success and Validity, and, among all the possible results that satisfy these two principles, it selects the one that has the minimal impact upon the original ontology (Principle of Minimal Change). Notice that there are certain limit cases which need to be handled separately, i.e, the case when the original ontology (to be updated) is invalid, and the case when the update itself is infeasible:

Definition 6. *Consider a language L, a validity model R, an update-generating ordering $<$ and an update algorithm $\bullet : L^+ \times L \mapsto L^+$. Then the algorithm $\bullet$ is called rational iff it satisfies the following requirements for all $O \subseteq L^+, U \subseteq L$:*

Limit Cases: *if O is not valid or U is infeasible, then $O \bullet U = O$*
General Case: *if O is valid and U is feasible, then $\bullet$ satisfies the following:*
  Principle of Success: $O \bullet U \vdash U$.
  Principle of Validity: $O \bullet U$ *is valid.*
  Principle of Minimal Change: *For any O' such that $O' \vdash U$ and O' is a valid ontology, it holds that $Delta(O, O \bullet U) \leq Delta(O, O')$.*

Note that rationality depends on the model (which determines L and L^+), the validity rules (for the Principle of Validity) and the ordering (for the Principle of Minimal Change). Therefore, there is no "universally rational update algorithm", but rationality depends critically on these parameters.

5 Algorithms

5.1 General-Purpose Algorithm

We will now show how one can use the above formal framework in order to develop a rational evolution algorithm (which is shown in Table 6). Let us consider the update example of Figure 3. Our original update is $U = \{C\_IsA(C, B)\}$, denoting that an IsA between C and B should be added. We first need to check whether this update will violate any rule (line 4.1); as mentioned in Section 4, this can be done by checking against all subrules in which $\neg C\_IsA$ appears. In general, several rules may be violated, in which case we process them in any order (line 4.2). In our example, it can be verified that the addition of $C\_IsA(C, B)$ will only violate subrule R12.1 (IsA transitivity), for $x = C, y = B, z = A$. This is true because the addition of $C\_IsA(C, B)$ should cause the addition of the implicit knowledge $C\_IsA(C, A)$ as well. This option is the standard way of satisfying transitivity, but our rule also gives us the alternative to remove the old IsA between B and A (to prevent the transitivity rule from firing).

In order to explore all alternatives regarding the possible side-effects, the comparison (using $<$) between the first and the second option is postponed until the full set of side-effects has been computed. Therefore, at this point, the algorithm suggests two different alternative updates, one per possible side-effect, namely $U_1 = \{C\_IsA(C, B), C\_IsA(C, A)\}$ and $U_2 = \{C\_IsA(C, B), \neg C\_IsA(B, A)\}$

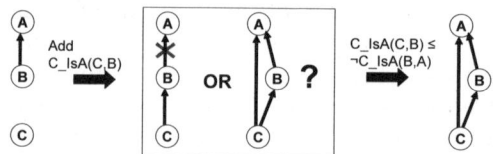

Fig. 3. Adding a class subsumption

(line 4.2.1). Then, the algorithm recursively calls itself twice (once for U_1 and once for U_2). Both calls will indicate no further side-effects, as there are no further rules violated; in the general case, the side-effects could have side-effects of their own, so the recursion should continue until no further side-effects exist. Once all recursions stop, the returned sets of side-effects are compared using $<$ and the minimal is selected for implementation (line 4.2.2). In this case, the first option (i.e., U_1) is the "best", according to $<$ (see Table 5), i.e., the IsA between C and A should be added; this indeed sounds like the most natural result, but it could be different if the ordering was different.

If, during the recursion, the so-far processed predicates turn out to contradict each other (line 1), then the particular branch of execution will obviously not lead to an acceptable solution, so the special value *infeasible* is returned; if all branches return *infeasible*, then the entire update is infeasible (and the recursive process will also return *infeasible*). The same special value is returned by certain branches (line 2) when their cost is predicted to be too large to be an acceptable solution, so there is no point in exploring them further.

Notice that the general algorithm (Table 6) is applicable for any language L (i.e., ontology model), validity model R and ordering $<$ and that several details of the algorithm have been brushed out. The general idea is that the case-based

Table 6. General-purpose algorithm

Input: Model, Rules, Ordering $<$, Update U, Ontology O
WHILE there exist unprocessed predicates in U execute the following steps:
(1) If the predicates that have been processed so far contradict each other, return INFEASIBLE
(2) If the total cost of the union of the predicates processed so far and the remaining predicates (in U) is larger than the best solution found so far, return INFEASIBLE
(3) Select (arbitrarily) an unprocessed predicate in U, say P
(4.1) IF there is no rule violated by P, THEN mark P as processed, add P to the side-effects of U and recursively call the algorithm using the same U
(4.2) ELSE select (arbitrarily) one violated rule, say R
(4.2.1) FOR each possible way to resolve the violation of R, add the respective predicates as side-effects in U and recursively call the algorithm using the new U
(4.2.2) When recursion returns compare (using $<$) the returned side-effects and return the "best" to the caller
Output: Update U enriched with its side-effects

---

reasoning performed manually in other systems is now in-built in the algorithm, so it is performed automatically and in a parameterizable way. The algorithm's complexity depends on its parameters, namely the language, validity model and ordering; for the particular parameters used for RDF (described above), termination can be guaranteed:

Theorem 1. *For the language, validity rules and ordering described in Section 4, the algorithm of Table 6 terminates for any input O, U.*

Termination is guaranteed by the form of the rules and the ordering (cost model) used. In particular, it can be shown that, whenever there exists a non-terminating recursive path (branch), there exists also a terminating one that is significantly less costly. By carefully choosing the processing order of the various side-effects (line 4.2.1), we can guarantee that the non-terminating branches will be pruned in line 2, before jeopardizing termination.

The algorithm described in Table 6 returns the effects and side-effects of the original update, or the special value *infeasible*. The end result of this recursive algorithm can then be trivially applied upon the original ontology, by simply adding every positive ground fact of the output to the ontology, and removing the positive counterpart of any negative ground fact of the output from the ontology. The result will be a valid ontology which should be returned as the result of the update. The following can be shown:

Theorem 2. *For any given language, validity rules and update-generating ordering, if the algorithm described above terminates, then it implements a rational change operation.*

The complete proof of the above theorem is quite complicated and technical, so we provide only a short sketch. Principle of Success is guaranteed by the fact that our algorithm considers all the predicates in U, and all such predicates are added to the side-effects of U (line 4.1). The Principle of Validity is guaranteed as well: the process cannot end unless all violated rules (identified in line 4.2) are restored (line 4.2.1). Finally, the Principle of Minimal Change is guaranteed in line 4.2.2: the recursive character of the algorithm will open up several different branches, each of them spawned by a different way to restore a particular rule violation. Upon returning of each branch, the calculated cost of each branch is compared (line 4.2.2) and only the best is kept; notice that the comparison is made at a position where the entire branch has been explored, so we know its total cost and can guarantee that no ignored branch can have "minimal" cost (so it can't be an acceptable solution). The following corollary is immediate:

Theorem 3. *For the language, validity rules and ordering described in Section 4, the algorithm of Table 6 terminates for any input $O \subseteq L^+, U \subseteq L$ and it can be used to implement a rational change operation.*

5.2   Special-Purpose Algorithms

A downside of the generality enjoyed by the algorithm of Table 6 is that it is not efficient. To remedy this problem, we can develop simpler, special-purpose algorithms, for the particular application that we are interested in (RDF in our case). These "instantiations" are much faster than the general algorithm, but can still be proven equivalent to it, i.e., formally sustained. Thus, we can guarantee that they exhibit the expected/desired behavior, by verifying them against the general-purpose algorithm. Notice that these special-purpose algorithms are similar to ad-hoc methodologies employed by other systems; however, using our formal framework and results, one can verify in a straightforward way the correctness (rationality) of those algorithms (see Theorem 4). Moreover, the general algorithm could still be used to implement any possible, unforseen operation.

Table 7 shows, as an example, one such special-purpose algorithm for the removal of a class from an ontology. Notice that some lines of the algorithm (e.g., (1.4.1)-(1.4.4)) would spawn other special-purpose algorithms for executing certain operations (in our case, the removal of IsAs, instantiation links etc), thus, possibly, incurring further side-effects. For this reason, similar algorithms have been developed for other operations, but are omitted due to space limitations.

Table 7. Special-purpose algorithm: remove class C from ontology O

Remove class C:
(1) If class C is in O THEN
(1.1) Remove all class IsA relationships deriving from C
(1.2) Remove all class IsA relationships arriving in C
(1.3) Remove all instantiation links between a resource and C
(1.4) FOR every property P whose range/domain is C
(1.4.1) Remove all property IsA relationships deriving from P
(1.4.2) Remove all property IsA relationships arriving in P
(1.4.3) Remove all instantiation links of P
(1.4.4) Remove P and the information on its range/domain
(1.5) Remove C

Theorem 4. *Consider the language, validity rules and ordering described in Section 4. Then for $U = \{\neg CS(C)\}$ and any $O \subseteq L^+$, the output of the algorithm in Table 6 is the same as the output of the algorithm in Table 7.*

The above theorem can be easily shown by exhaustively considering all the different rule violations that the update under question would cause (by scanning the validity rules for violations); this would verify that the behavior of the special-purpose algorithm is identical to the general-purpose one for the particular order considered. Similarly to the other ontology evolution systems, our special-purpose algorithms cannot handle all possible update requests. However, we can always resort to the general-purpose algorithm if the requested operation is not supported by any special-purpose algorithm. Currently, we have devised and implemented one special-purpose algorithm for each elementary operation, but we plan to develop more, in order to handle certain useful composite operations. The selection whether to use a special-purpose algorithm or the general-purpose one is made by the system itself, in a transparent manner to the user.

6 Conclusion

In this paper, we identified several difficulties associated with the development of ad-hoc ontology evolution algorithms. We decomposed the process of coping with ontology evolution into 5 discrete steps. This way, devising an ontology evolution algorithm is reduced to the process of instantiating each step in a modular way. To this end, we presented a formal framework with the aid of which an evolution algorithm can be materialized as a set of adequate parameterizations, as follows:

1. The ontology representation model and its mapping to FOL.
2. The definition of the allowed change operations in the model. Notice that this is not necessary, as the framework is general enough to support any update, but we may want to disallow certain operations for some application.
3. The validity rules that allow us to detect invalidities as well as to determine how the invalidities can be resolved.
4. The preference ordering that encodes the selection mechanism.

Parameters 1,2 and 4 of our framework correspond to steps 1,2 and 5 respectively. The third parameter corresponds to the validity context, based on which our framework instantiates steps 3 and 4. Once these parameters are set, we can apply the general algorithm presented in Table 6 to perform any change. For efficiency reasons, it may be useful to generate simpler special-purpose algorithms based on the general one. This can be done only for specific instantiations of the above parameters, as in the case study of RDF updating presented here.

Our method exhibits a faithful behavior with respect to the various choices involved, regardless of the particular ontology or update operation at hand. It has a formal foundation, issuing a solid, consistent and customizable method to handle any type of change operation, including updates that have not been considered at design time. Our framework is modular and extensible in the sense that it could work with any language, rules and/or ordering given.

As already mentioned, the presented algorithms have been implemented for the Change Impact Service of SWKM, and the initial results are promising. In the future, we plan to identify and optimize the most commonly used update operations. In addition, we plan to verify the effectiveness of our proposed ordering using experiments with real users.

Acknowledgements

This work was partially supported by the EU projects CASPAR (FP6-2005-IST-033572) and KP-Lab (FP6-2004-IST-4).

References

1. Bechhofer, S., Horrocks, I., Goble, C., Stevens, R.: OilEd: a Reason-able Ontology Editor for the Semantic Web. In: Baader, F., Brewka, G., Eiter, T. (eds.) KI 2001. LNCS (LNAI), vol. 2174. Springer, Heidelberg (2001)

2. Benedikt, M., Bruns, G., Gibson, J., Kuss, R., Ng, A.: Automated update management for XML integrity constraints. In: Proc. Workshop on Programming Languages for XML (PLAN-X) (2002)
3. Dalal, M.: Investigations Into a Theory of Knowledge Base Revision: Preliminary Report. In: Proceedings of the 7th National Conference on Artificial Intelligence (AAAI 1988), pp. 475–479 (1988)
4. Flouris, G.: On the Evolution of Ontological Signatures. In: Proceedings of the Workshop on Ontology Evolution (OnE 2007) (2007)
5. Flouris, G., Manakanatas, D., Kondylakis, H., Plexousakis, D., Antoniou, G.: Ontology Change: Classification and Survey. Knowledge Engineering Review (KER) (to appear)
6. Gabel, T., Sure, Y., Voelker, J.: KAON-Ontology Management Infrastructure. SEKT informal deliverable 3(1) (2004)
7. Gärdenfors, P.: Belief Revision: An Introduction. In: Gärdenfors, P. (ed.) Belief Revision, pp. 1–20. Cambridge University Press, Cambridge (1992)
8. Laurent, D., Phan Luong, V., Spyratos, N.: Updating intensional predicates in deductive databases. Data & Knowledge Engineering 26(1), 37–70 (1998)
9. Munoz, S., Perez, J., Gutierrez, C.: Minimal Deductive Systems for RDF. In: Franconi, E., Kifer, M., May, W. (eds.) ESWC 2007. LNCS, vol. 4519. Springer, Heidelberg (2007)
10. Noy, N., Fergerson, R., Musen, M.: The Knowledge Model of Protégé-2000: Combining Interoperability and Flexibility. In: Dieng, R., Corby, O. (eds.) EKAW 2000. LNCS (LNAI), vol. 1937, pp. 17–32. Springer, Heidelberg (2000)
11. Noy, N., Klein, M.: Ontology Evolution: Not the Same as Schema Evolution. Knowledge and Information Systems 6(4), 428–440 (2004); also available as SMI technical report SMI-2002-0926 (2004)
12. Qi, G., Liu, W., Bell, D.A.: Knowledge Base Revision in Description Logics. In: Fisher, M., van der Hoek, W., Konev, B., Lisitsa, A. (eds.) JELIA 2006. LNCS (LNAI), vol. 4160, pp. 386–398. Springer, Heidelberg (2006)
13. Serfiotis, G., Koffina, I., Christophides, V., Tannen, V.: Containment and Minimization of RDF/S Query Patterns. In: Gil, Y., Motta, E., Benjamins, V.R., Musen, M.A. (eds.) ISWC 2005. LNCS, vol. 3729. Springer, Heidelberg (2005)
14. Stojanovic, L., Maedche, A., Motik, B., Stojanovic, N.: User-driven Ontology Evolution Management. In: Gómez-Pérez, A., Benjamins, V.R. (eds.) EKAW 2002. LNCS (LNAI), vol. 2473. Springer, Heidelberg (2002)
15. Stojanovic, L., Motik, B.: Ontology Evolution Within Ontology Editors. In: Proceedings of the OntoWeb-SIG3 Workshop, pp. 53–62 (2002)
16. Stuckenschmidt, H., Klein, M.: Integrity and Change in Modular Ontologies. In: Proceedings of the 18th International Joint Conference on Artificial Intelligence (IJCAI 2003) (2003)
17. Sure, Y., Erdmann, M., Angele, J., Staab, S., Studer, R., Wenke, D.: OntoEdit: Collaborative Ontology Development for the Semantic Web. In: Horrocks, I., Hendler, J. (eds.) ISWC 2002. LNCS, vol. 2342. Springer, Heidelberg (2002)

Relational Databases in RDF:
Keys and Foreign Keys

Georg Lausen

University of Freiburg, Institute for Computer Science
Georges-Köhler-Allee, 79110 Freiburg, Germany
`lausen@informatik.uni-freiburg.de`

Abstract. Today, most of the data on the web resides in relational
databases. To make the data available for the semantic web mappings
into RDF can be used. Such mappings should preserve the information
about the structure of keys and foreign keys, because otherwise important semantic information is lost. In this paper, we discuss several possible ways to map relational databases into an RDF graph. We discuss
the problem of how to represent the original key and foreign key constraints in the resulting RDF graph and demonstrate, that different kinds
of mappings require different solutions. We finally propose to explicitly
represent the structure of keys and foreign keys by means of the vocabulary of a new RDF namespace.

1 Introduction

Today, most of the data on the web resides in relational databases. Typically,
even in the case when several databases represent information about the same application domain, they use different schemata. The idea of the semantic web is to
support semantic interoperability between programs exchanging data. To achieve
this goal W3C has standardized several languages to define ontologies [17], which
can be used to define common vocabularies and structures for certain applications. Prominent examples are the *Resource Description Framework* RDF [21],
which is considered as the basis for building the semantic web, and the *Web
Ontology Language* OWL, whose mostly considered variant is based on a Description Logic [18,1]. Using RDF, any kind of information can be represented
by a set of so called triples, where each triple states a subject-property-object
relationship. As each such triple can be understood as a directed edge from the
subject to the object, where the edge is labelled with the respective property,
instead of a set of triples a corresponding RDF graph is considered. Thus, ex-
porting data from relational databases to the semantic web using RDF basically
means to map the relational data into an RDF graph.

A relational database may be physically exported as an RDF graph according
to some mapping, respectively mapped into RDF in a similar fashion to a virtual relational view definition. Several approaches to define such mappings have
already been described [2,5,4,12], however much less attention has been payed
on the question how key and foreign key expressions stated in the relational

V. Christophides et al. (Eds.): SWDB-ODBIS 2007, LNCS 5005, pp. 43–56, 2008.

database can be represented in RDF. In [12,8] a mapping is proposed which aims at the representation of keys and foreign keys as well, however the proposed mapping is specific to mapping an Entity-Relationship Model into RDF.

If key and foreign key information is lost when mapping a relational database into an RDF graph, the representational quality of the result has degraded substantially. This may become an important issue,

- if a user builds her own knowledge base by integrating several RDF graphs found on the internet,
- if an exported RDF graph is imported in a relational database at another place,
- if updates on a materialized RDF graph have to be performed such that key and foreign key properties have to be checked.

Finally, if an RDF graph is upgraded to OWL, keys and foreign keys could become interesting for reasoning, as well.

In this paper we will discuss various ways to map a relational database into RDF. In principle, many different methods may be applied. For each such mapping we will discuss how key and foreign key constraints can be expressed. For some mappings this will be rather straightforward, while other mappings may require additional constraints, or will require rather contrived solutions. Complications will arise in situations where a key is built out of several attributes and some of them are used for a foreign key as well. In addition, RDF is proposed as a simple data model which allows anyone to make statements about any resource. Thus a mapping should not only allow to represent the constraints, however has to give credit to the philosophy of RDF as well. Based on these observations we consider the mapping problem from relational databases in RDF still to be not sufficiently understood.

As key and foreign key constraints can not be asserted inside RDF, we will finally argue that RDF should be extended by a vocabulary which allows the declaration of keys and foreign keys. When keys and foreign keys are explicitly stated as part of an RDF graph, RDF processors are enabled to check the corresponding constraints and in this way are able to guarantee important quality criteria of RDF graphs.

The structure of the paper is as follows. In Section 2 we shall introduce the basic formalism and a running examples which we will use to demonstrate the various mappings. Mappings from relational databases into RDF are presented and discussed in Section 3. In Section 4 we present a vocabulary which allows to express key and foreign key constraints inside an RDF graph. Section 5 finally concludes the paper.

2   Preliminaries

We shall first introduce the terminology we use in the sequel. We start with relational databases (cf. [10], Section 3). A relational database schema **R** is a set of relation schemata identified by R, $\mathbf{R} = (R_1, ..., R_n)$. We use $Att(R)$ to denote the set of attributes of the relation symbol R. An instance $\mathbf{I}$ of $\mathbf{R}$ is a tuple $(I_1, ..., I_n)$, where for $1 \leq i \leq n$ I_i is a finite instance of R_i, i.e. a finite subset of the n-ary cartesian product over an underlying domain. An element $\mu \in I$ is called tuple. Let $A \in Att(R)$, we use $\mu.A$ to denote the value of the attribute A of the tuple μ.

A *key* over $\mathbf{R}$ is an expression of the form $R[A_1, ..., A_k] \rightarrow R$, where $R \in \mathbf{R}$ and for $1 \leq i \leq k$ it holds that $A_i \in Att(R)$.[1] Let $\mathbf{I}$ be an instance of $\mathbf{R}$. $\mathbf{I}$ satisfies $R[A_1, ..., A_k] \rightarrow R$ if and only if $\forall \mu_1, \mu_2 \in I$ $(\bigwedge_{1 \leq i \leq k}(\mu_1.A_i = \mu_2.A_i) \rightarrow \bigwedge_{A \in Att(R)}(\mu_1.A = \mu_2.A))$.

A *foreign key* over $\mathbf{R}$ is an expression of the form $R[A_1, ..., A_k] \subseteq R'[A'_1, ..., A'_k]$, where $R, R' \in \mathbf{R}$, $\{A_1, ..., A_k\} \subseteq Att(R)$, $\{A'_1, ..., A'_k\} \subseteq Att(R')$, and $R'[A'_1, ..., A'_k] \rightarrow R'$. R is called *child* and R' *parent* of the foreign key. $\mathbf{I}$ satisfies $R[A_1, ..., A_k] \subseteq R'[A'_1, ..., A'_k]$ if and only if $\mathbf{I}$ satisfies $R'[A'_1, ..., A'_k] \rightarrow R'$ and $\forall \mu_1 \in I$ $\exists \mu_2 \in I'$ $(\bigwedge_{1 \leq i \leq k} \mu_1.A_i = \mu_2.A'_i)$.

Next we introduce the required RDF terminology adapting definitions in [20] for our purposes. We consider a *vocabulary* $\mathcal{V} = (N_C, N_P)$, where N_C is a finite set of *classes* and N_P is a finite set of *properties*. Given a vocabulary $\mathcal{V}$, an *interpretation* $\mathcal{I} = (\Delta_I, \Delta_D, .^{I_C}, .^{I_P})$ of $\mathcal{V}$ is given as follows:

- Δ_I is a nonempty set, called *object domain*,
- Δ_D is a nonempty set, called the *data domain*, which we assume to be disjoint from Δ_I, $\Delta_I \cap \Delta_D = \emptyset$,
- $.^{I_C}$ is the *class interpretation function* assigning to each class $C \in N_C$ a finite subset $C^{I_C} \subseteq \Delta_I$,
- $.^{I_P}$ is the *property interpretation function* assigning to each property $Q \in N_P$ a finite subset $Q^{I_P} \subseteq \Delta_I \times (\Delta_I \cup \Delta_D)$.

Based on a given interpretation we can introduce a corresponding RDF graph. In the RDF document [21] among the nodes of an RDF graph it is distinguished between RDF URI references and literals. Literals in our framework are the elements of the data domain. By requiring that the object domain and the data domain are disjoint, we assume that literal values are not identified by URIs [22]. Therefore, in an RDF graph, literals with identical value will be represented by different nodes. This fact makes the following definitions a bit more complicated.

Let $\mathcal{I} = (\Delta_I, \Delta_D, .^{I_C}, .^{I_P})$ be an interpretation. The *RDF graph* $G^I = (N^I, E^I)$ of $\mathcal{I}$ then is a directed labelled graph, where

- for the set of nodes N^I we have $N^I = N_C \cup \{a, b \mid (a, b) \in Q^{I_P}, Q \in N_P, b \in \Delta_I\} \cup \{a, b_{a,Q} \mid (a, b) \in Q^{I_P}, Q \in N_P, b \in \Delta_D\}$, and
- for the set of labelled arcs E^I we have $E^I = \{(a, Q, b) \mid (a, b) \in Q^{I_P}, b \in \Delta_I\} \cup \{(a, Q, b_{a,Q}) \mid (a, b) \in Q^{I_P}, b \in \Delta_D\} \cup \{(a, \mathtt{rdf:type}, C) \mid a \in C^{I_C}, C \in N_C\}$.

---

[1] We consider for each schema only one key, say the primary key.

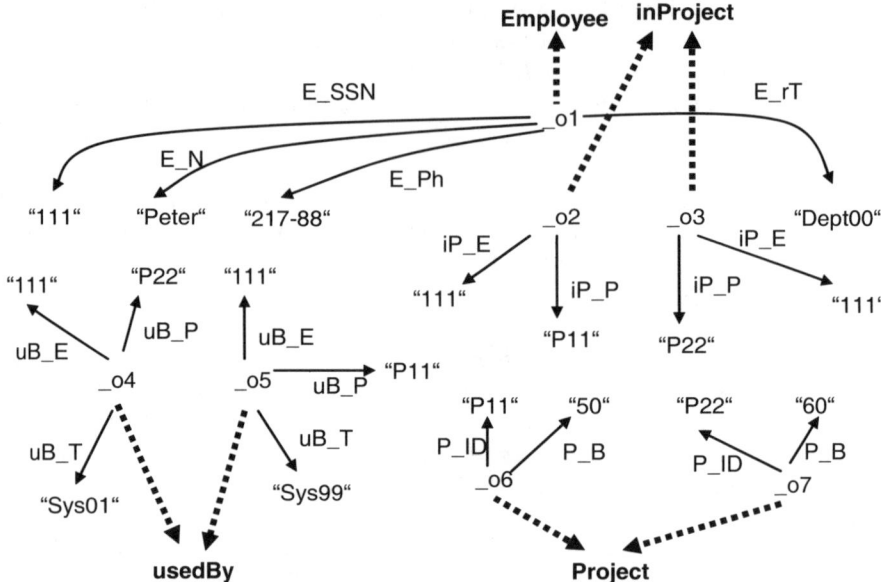

Fig. 1. Tuple-based mapping. Classes are indicated by bold font and typing by dotted edges.

We exemplify our discussion by means of a small running example. Consider the following relational schemata:

```
Employee(SSN, Name, Phone, reportsTo)
Project(ID, Budget)
inProject(Empl, Proj)
usedBy(Tool, Proj, Empl)
```

To indicate the key of a schema we use bold font. The schemata represent a scenario in which employees are described by name, ID, phone, and by the department they report to. Employees may be involved in projects and employees may use tools, where for a certain tool with respect to a certain project there is at most one employee who is using it.

We observe that Empl and Proj are foreign keys in schema inProject. Moreover, Empl and Proj together form a foreign key in schema usedBy with respect to inProject. For concreteness, we consider the following instances:

Employee				Project	
SSN	Name	Phone	reportsTo	**ID**	Budget
111	Peter	217-88	Dept00	P11	50
				P22	60

inProject		usedBy		
Empl	**Proj**	**Tool**	**Proj**	Empl
111	P11	Sys99	P11	111
111	P22	Sys01	P22	111

A simple kind of mapping from a relational database into RDF is as follows. Let R be a relation schema, where $Att(R) = \{A_1, \ldots, A_k\}$. Introduce a class C_R and properties $Q_{R,A_1}, \ldots, Q_{R,A_k}$. Let I be an instance of R. For every tuple $\mu = (a_1, \ldots, a_k)$ in I introduce a unique blank node $\_n_\mu$ and a labelled edge $(\_n_\mu, \text{rdf} : \text{type}, C_R)$. The naming of such blank nodes is arbitrary; for simplicity we can think of a numbering scheme based on incrementing a counter to assign to every blank node a unique number. For every nonnull value $\mu.A$ of μ, $A \in Att(R)$, introduce an edge $(\_n_\mu, Q_{R,A}, (\mu.A)\_{n_\mu,Q_{R,A}})$.[2] This mapping treats all tuples in the relation instances separately; we shall call the mapping *tuple-based*. Applying the tuple-based mapping we will get the RDF graph depicted in Figure 1 in which all property values are taken from the data domain, i.e. represented by literals in quotation marks. This mapping totally ignores that foreign key values also appear as key values and therefore refer to the same object. A tuple-based mapping thus has the following two severe deficiencies. It produces a lot of redundancies as foreign key values appear separate to key values and there is nothing in the graph which explicitly relates objects according to a foreign key relationship. Therefore, tuple-based mappings will not be considered further. The mappings we shall introduce in the next section demonstrate that these deficiencies result from an imperfect modelling and that they are not inherent to RDF.

3   Mapping Relational Databases into RDF

There have been several approaches described in the literature, which elaborate on the mapping of relational databases to an RDF graph (e.g. [2,5,4,12]). In this section we will discuss mappings with an eye on constraints. We will analyze possible ways to express the relational key and foreign key constraints in the resulting RDF graph. The following definitions clarify our notions of relational key and foreign key in the context of RDF.

To define a key constraint for RDF we first require that each property is interpreted by a (total) function. Thus, whenever a property $R \in N_P$ is involved in a key constraint, we require an additional functionality constraint to be given. Let $Q \in N_P$. We write $Func(C, Q)$ to state that for any object $o \in C$, on which a property Q is defined, Q associates o with exactly one other object. We write $Key(C, Q_1, \ldots, Q_n)$ to state a relational key of class C over properties $Q_1, \ldots, Q_n$ and we write $FK(C, [Q_1, \ldots, Q_n], C', [Q'_1, \ldots, Q'_n])$ to state a relational foreign key over the respective properties of child C and parent C'.

Definition 1. *Let* $\mathcal{I} = (\Delta_I, \Delta_D, .^{I_C}, .^{I_{P_o}})$ *be an interpretation. Let* $C \in N_C$, $Q, Q_i, Q'_i \in N_P$, $1 \leq i \leq n$. *Let* ψ *be a constraint.* $\mathcal{I}$ *satisfies* ψ, $\mathcal{I} \models \psi$, *if there holds:*

– *Let* ψ *be a functionality constraint* $Func(C, Q)$.
  $$\{x \mid \#\{y \mid (x, y) \in Q^{I_P}\} \neq 1, x \in C^{I_C}\} = \emptyset.$$

---

[2] A discussion how to represent a null value in RDF is beyond the scope of this paper.

- Let ψ be a relational key constraint $Key(C, Q_1, \ldots, Q_n)$ and let $\mathcal{I} \models Func(C, Q_i), 1 \leq i \leq n$.
  If $\exists o_1, o_2 \in C^{\mathcal{I}_C}$ such that $\exists v_i \in \Delta_I \cup \Delta_D, 1 \leq i \leq n$,
    where $(o_1, v_i), (o_2, v_i) \in Q_i^{I_P}$, then $o_1 = o_2$.
- Let ψ be a relational foreign key constraint $FK(C, [Q_1, \ldots, Q_n], C', [Q'_1, \ldots, Q'_n])$ and let $\mathcal{I} \models Key(C', Q'_1, \ldots, Q'_n), 1 \leq i \leq n$.
  If $o_1 \in C^{\mathcal{I}_C}$ then $\exists o_2 \in C'^{\mathcal{I}_C}$ such that
    $(o_1, v_i) \in Q_i^{I_P}$ implies $(o_2, v_i) \in Q_i'^{I_P}, 1 \leq i \leq n$.

Keys and foreign keys according to these definitions are called *relational* to distinguish them from more generally defined keys. It is required that objects can be uniquely identified by properties directly related to them. Keys in a more general sense [14] could also be defined by means of path expressions in which the uniquely identifying property is defined by the composition of functional properties. We will return to this issue later again.

3.1   Value-Based Mappings

There are several ways to improve the tuple-based mapping. As a first approach, we represent key attribute values by URIs[3] and now are able to reference the

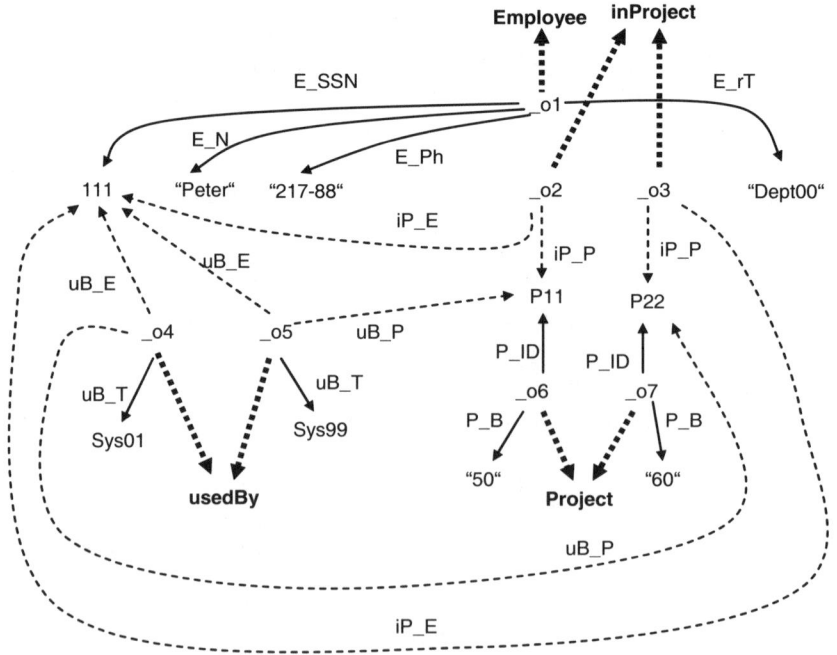

Fig. 2. Value-based mapping. Values of key attributes are URIs. References to foreign key attribute values are represented by dashed edges.

---

[3] For space reasons and not to overload our notation we consider only relative URIs in this paper.

key attribute values of parent objects from the respective child objects. This is demonstrated by Figure 2. Mappings following this approach are called *value-based* mappings. For example, properties IP_P and iP_E of the objects in class inProject now directly refer to the key values of the respective objects of class Employee and Project. Analogously, with respect to the foreign keys of the objects in class usedBy, we now directly refer to the respective key values of the objects in class Employee and Project.

Value-based mappings preserve the key and foreign key constraints that have been defined in the source relational database, because every key and foreign key stated in the relational database can be expressed by a relational key and foreign key constraint over the resulting RDF graph. We will next show, that such a direct correspondence does not immediately hold for other mappings, whose application seems to be quite natural in the context of RDF.

3.2 URI-Based Mappings

A popular mapping approach is based on encoding key values in URIs such that each object can be uniquely identified (cf. [4,5]) and then be further described by its properties. RDF triples that express links between such URIs express nothing else than foreign key relationships. At a first glance such an approach seems to solve the representation problem of keys and foreign keys in RDF in a very elegant and concise way.

However, keys may be built out of several attributes which themselves may be even foreign key attributes, in general. This typically will happen when relations are used to represent relationships between objects. In the Employee-Project-inProject example, the key of relation inProject is given by the keys of relations Employee and Project; we thus assume that one employee may work for several projects and one project may be processed by several employees. Having formed the corresponding URIs out of key values, the original foreign key constraint now appears as a less explicit constraint on the URI string-values. For example, if employee 111 works for project P11, then this will give rise to a relationship between 111 and P11. Following the above sketched approach we would first introduce the two URIs 111, P11, then a combined URI 111&P11 and then define triples (111&P11, iP_E, 111), respectively (111&P11, iP_P, P11). As we can see, the foreign key problem still exists; however the foreign key relationships cannot be expressed over triples, but over URI values. This means, whenever there exists a URI 111&P11 in the RDF graph, there must also exist a triple (111&P11, iP_E, 111) and we would not accept a triple (111&P11, iP_E, 222). In the RDF standard [21] RDF is proposed as an open-world framework that allows anyone to make statements about any object (resource). If a user is going to make a statement about object 111&P11, she would certainly expect that there exists an object 111 as well. This means that there are still constraints reflecting foreign key constraints which are to be guaranteed.

The technique we have just described is one of the mappings suggested in [4,5], where such URIs are formed using string concatenation. However the syntax of URIs gives us enough room to construct URIs from key values in a more

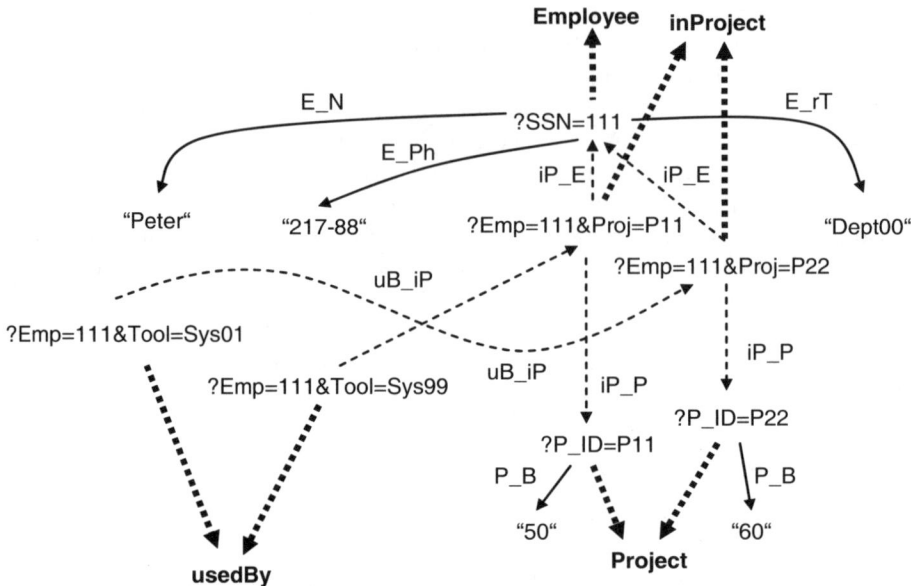

Fig. 3. URI-based mapping. Objects are represented by URIs which are build out of the corresponding key values. References between objects are represented by dashed edges.

standardized way. We propose to use the query part of an URI to build unique URIs by still making the origin of the components of such URIs visible. This is demonstrated in Figure 3. We call mappings following this idea *URI-based* mappings.

The discussion so far clearly shows that using key values to construct URIs representing an object may produce complications even though the approach looks attractive at first glance. When such URIs contain foreign key values, then a consistency problem arises which can only be solved by carefully checking the string representations of the URIs. For example, with respect to URI ?Emp=111&Tool=SYS99 representing an object of class usedBy and URI ?Emp=111&Proj=P11 representing an object of class inProject it has to be guaranteed that both Emp-parts are equal.

3.3   Object-Based Mappings

The next type of mappings identify objets by blank nodes and represent foreign key relationships by links between such blank nodes. We call this kind of mapping *object-based*; such mappings are also possible using the approach proposed in [4]. Different to the URI-based mappings, keys still are explicitly represented by properties. Therefore, in principle, key values can be represented by literals. However, as we will see later, there are reasons why it should be not only possible to reference objects, but key values as well. Therefore, key values are represented

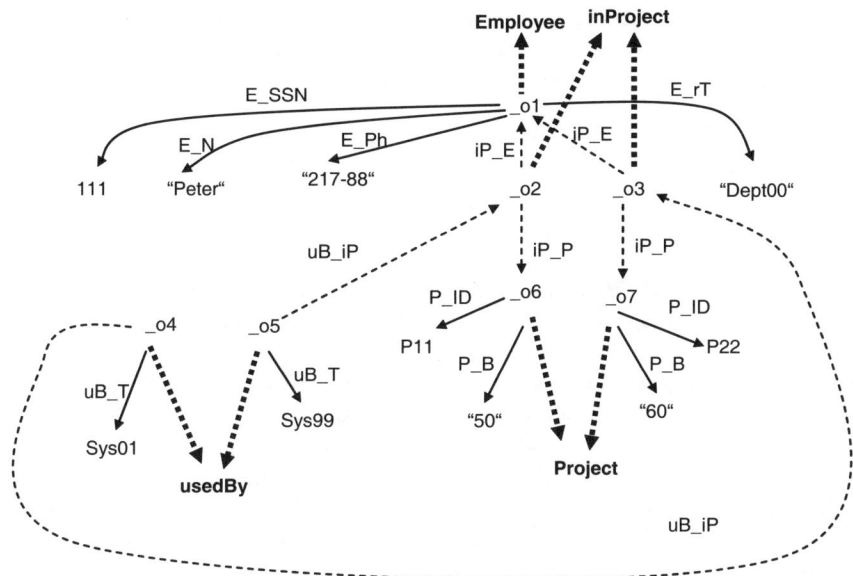

Fig. 4. Object-based mapping

by URIs in the sequel. As with the URI-based mappings there is no need to explicitly introduce properties for foreign keys - a link from the child object to its parent is sufficient. A mapping of this kind is shown in Figure 4.

As we can see, foreign key constraints are represented by links. However we need a constraint to ensures that such links will indeed exist. Such constraints are called *participation* constraints and are well-known from entity-relationship modelling (see also [15]). Let $C, C' \in N_C$ be classes and $Q \in N_P$ a property. Let Q represent a foreign key relationship and consider C as the child class and C' the parent class. We write $Partcipate(C, Q, C')$ to state a participation constraint which can be formalized in our RDF setting as follows.

Definition 2. *Let* $\mathcal{I} = (\Delta_I, \Delta_D, .^{I_C}, .^{I_{Po}})$ *be an interpretation. Let* $\phi = Partcipate(C, Q, C')$. *We define* $\mathcal{I}$ *satisfies* ϕ, $\mathcal{I} \models \phi$, *if there holds:*

$$\text{If } o_1 \in C^{I_C} \text{ then } \exists o_2 \in C'^{I_C} \text{ such that } (o_1, o_2) \in Q^{I_P}.$$

When carefully looking at Figure 4 the question arises where we have represented the key values of the objects of class usedBy with respect to the property giving us the ID of the project for which a certain employee uses a tool. As the key and the foreign key of class usedBy overlap, in some sense, the foreign key has stolen the key property giving us the project ID being part of the key. Therefore, the key for usedBy is not given by properties directly associated with usedBy, but given by property uB_T of usedBy and the path formed by the functional properties uB_iP, iP_P and P_ID. Thus, for being able to check key constraints, the notion of relational keys is too limited and a more general definition of key is in order. We write $Key(C, Q_1^\circ \ldots Q_n^\circ)$ to state a general key constraint of class C over property paths $Q_1^\circ, \ldots, Q_n^\circ$, where each Q_i°, $1 \leq i \leq n$ is a path of functional properties $Q_{ij} \in N_P$, $1 \leq i \leq n$, $1 \leq j \leq n_i$, $n_i \geq 1$, which is written $Q_{i1} \ldots Q_{in_i}$.

Definition 3. *Let* $\mathcal{I} = (\Delta_I, \Delta_D, \cdot^{I_C}, \cdot^{I_{P\circ}})$ *be an interpretation. Let* $C \in N_C$, Q_i° *a path of functional properties* $Q_{i1}, \ldots, Q_{in_i}$, *where* $Q_{ij} \in N_P$, $1 \leq i \leq n$, $1 \leq j \leq n_i$, $n_i \geq 1$. *Let* ψ *be a* general key constraint $Key(C, Q_1^\circ, \ldots, Q_n^\circ)$. $\mathcal{I}$ *satisfies* ψ, $\mathcal{I} \models \psi$, *if there holds:*[4]

$$\text{If } \exists o_1, o_2 \in C^{I_C} \text{ such that } \exists v_i \in \Delta_I \cup \Delta_D, 1 \leq i \leq n,$$
$$\text{where } (o_1, v_i), (o_2, v_i) \in Q_{i1}^{I_P} \circ \ldots \circ Q_{in_i}^{I_P}, \text{ then } o_1 = o_2.$$

General keys have been studied in description logics [14] and object-oriented databases [6,3]; the key concept as proposed by XML Schema [24] resembles general keys, as well. If we understand RDF as a formalism which has been introduced to allow to extend a given RDF graph by new information in a very flexible and open way, it is not clear to us, whether it is feasible to allow to define key properties by paths of properties. Keys seem to us much easier to comprehend when they are directly related with the respective objects. In Figure 5, to this end we introduce a property uB_P which applied on objects of class usedBy returns the missing P_ID-value such that the key Tool, Proj for objects of class usedBy now can be expressed by a relational key again. The price we have to pay is an additional constraint which guarantees that property uB_P and property path uB_iP.iP_P.P_ID will give the same value when applied on the same object. Such constraints have been called *subobjectproperty-chain* [19]. We write $SubPChain(C, Q_1, \ldots, Q_n, S)$ to state a subobjectproperty-chain and formalize it according to our needs as follows:

Definition 4. *Let* $\mathcal{I} = (\Delta_I, \Delta_D, \cdot^{I_C}, \cdot^{I_{P\circ}})$ *be an interpretation. Let* $\phi = SubPChain(C, R_1, \ldots, R_n, S)$. *We define* $\mathcal{I}$ *satisfies* ϕ, $\mathcal{I} \models \phi$, *if there holds:*[5]

$$\{(x, y) \mid (x, y) \in R_1^{I_P} \circ \ldots \circ R_n^{I_P}, x \in C^{I_C}\} =$$
$$\{(x, y) \mid (x, y) \in S^{I_P}, x \in C^{I_C}\}.$$

3.4   Discussion

In the preceding sections we have discussed various kinds of mappings from relational databases to RDF. This discussion does not claim to have been exhaustive. For example, [12] proposes a mapping in which attributes A are represented by objects o_A and each tuple t of a (entity-) relation identifies a property Q_t which then is used to assign o_A some value v out of the object or value domain. This would give rise to the triple (o_A, Q_t, v). This technique is interesting as well and is based on the original formal definitions of the Entity-Relationship Model [9].

---

[4] The operator $\circ$ denotes the composition of binary relations.

[5] In contrast to [19] we can require equality of sets instead of a subset-relationship, because all properties involved are functional.

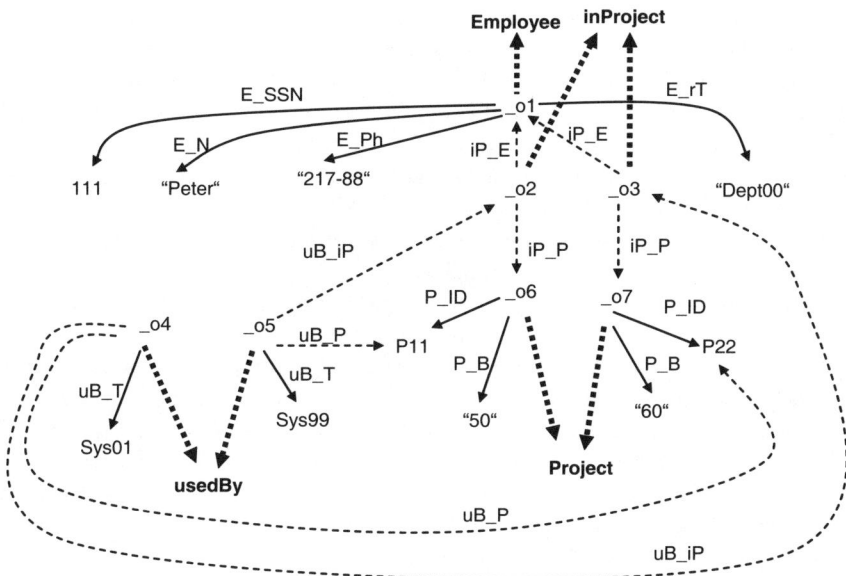

Fig. 5. Object-based mapping. To directly relate each object with its key properties, additional properties are introduced.

However, values may also appear as subjects in other triples whose predicates now do not correspond to tuples in a (entity-) relation, however to attributes of the given Entity-Relationship schema. Therefore, to our understanding, a resulting RDF graph is likely to be difficult to comprehend.

When mapping relational databases into RDF, we can choose mappings which preserve relational constraints and still use the linking capabilities of RDF (value-based mapping). However, we can also choose mappings which try to take the most advantage of the flexibility offered by RDF. Such mappings are attractive, but make it rather contrived to express key and foreign key constraints (URI-based mappings), or enforce the introduction of other kinds of constraints, which are not familiar in relational databases (object-based mapping). Which approach is the most appropriate one depends on the kind of applications for which a resulting RDF graph is assumed to be used. In the next section we will outline a technique that will allow us to state relational key and foreign key constraints as part of an RDF graph. A value-based mapping in conjunction with an explicit statement of the existing key and foreign key constraints inside the same RDF graph seems to us an attractive compromise between relational modelling and RDF modelling.

4 Keys and Foreign Keys Vocabulary for RDF

To get a clean and general solution for expressing keys and foreign keys inside an RDF graph we now propose to make that information explicit by means of an appropriate namespace. As a first idea one could use the XML-schema namespace [24], which allows to express keys and foreign keys, where the latter are called keyref. However, XML in contrast to RDF is a hierarchical model equipped with an implicit notion of order. Therefore, semantics of keys and foreign keys in XML and RDF are different, though similar in spirit. For these reasons we decided to extend the RDF vocabulary by a new dedicated namespace, which will be identified by prefix rdfc. This namespace extends the RDF vocabulary by two (meta-) classes rdfc:Key and rdfc:FKey, whose instances represent the key and foreign key definitions of the (application) classes. In addition, the namespace contains properties rdfc:Key, rdfc:FKey and rdfc:Ref which will allow to attach keys and foreign keys to classes and to associate foreign keys with the respective key of the parent class. The construction we shall outline is in the spirit of the PRIMARY KEY and REFERENCE-clause in SQL.

Keys and foreign keys may be built out of several components and therefore will be of type rdf:Bag. The construction is demonstrated in Figure 6. Starting from the graph depicted in Figure 2 we proceed as follows. We introduce objects E_Key, P_Key, iP_Key and uB_Key, respectively. In addition we introduce objects iP_FKey1, iP_FKey2 and uB_FKey to represent the two foreign keys of class inProject and the foreign key of class usedBy. We add corresponding typing edges from the newly introduced objects to classes rdfc:Key and class rdfc:FKey, respectively. Next, we add the corresponding edges labelled with

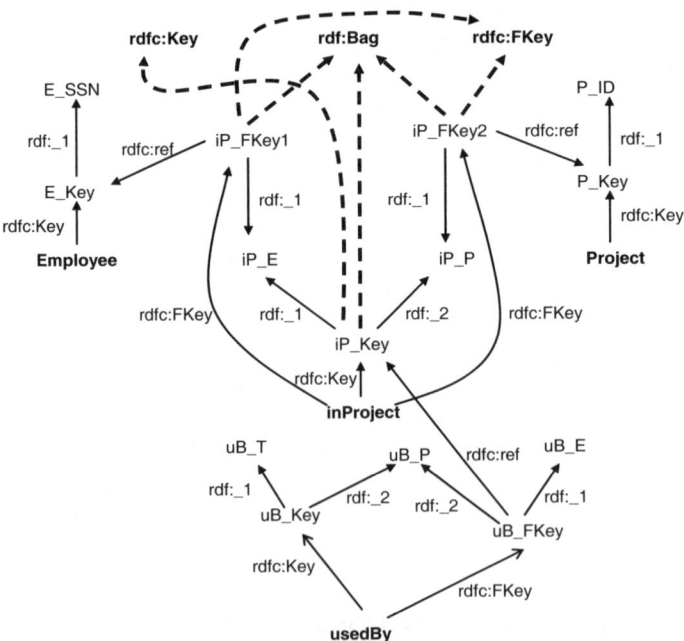

Fig. 6. Keys and foreign keys definitions are explicitly represented as part of an RDF graph. The typing of keys and foreign keys is visualized only for class inProject.

`rdfc:Key`, respectively `rdfc:FKey`, in order to relate a key and foreign key to its respective class, i.e. one class out of `Employee, Project, inProject, usedBy` for each key, respectively foreign key. In addition, for each foreign key we introduce an edge labelled `rdfc:ref` to associate the foreign key with the respective key of the parent class. Finally, as keys and foreign keys are of type `rdf:Bag` as well, we add corresponding typing edges and also indicate the components of keys and foreign keys by the edges labelled with `rdf:_1` and `rdf:_2`.

If we assume an RDF query language be given, which allows to traverse a RDF graph in edge and inverse edge direction, e.g. SPARQL [23], it is easy to see, how for each class its key and its foreign keys can be determined. Therefore, whenever we are interested in this information, e.g. for checking the constraints, we can first extract the constraints from the graph and then perform the checking. In [13] we show that all these steps can be performed using SPARQL.

5 Conclusion

In this paper we have argued to introduce key and foreign key constraints into RDF to explicitly state semantic information about the information represented by an RDF graph. Checking of the constraints is possible by testing appropriate SPARQL-queries for emptiness (cf. [13]), in a similar way to constraint checking using SQL. In [15,16,7] constraints are investigated with respect to OWL. In [15] the fundamental difference between constraints in relational databases and ontology languages is discussed and [16,7] present possible ways of integration. While these approaches emphasize the description logic point of view on constraints, we are interested in the question, how key and foreign key constraints can be expressed depending on the particular mapping applied from a relational database into an RDF graph. In addition, we show how these constraints can be expressed as part of an RDF graph. As a consequence, processing of the RDF data originating from relational data and RDF data originating from relational schema (meta-) data, in particular keys and foreign keys, becomes possible, e.g. using SPARQL or rule languages like F-Logic [11]. In contrast, the description logic based approaches allow query answering and powerful logical reasoning. Our approach to extend RDF by constraints and the description logic based approach can therefore be considered orthogonal and we expect that both will be useful for semantic web applications.

Acknowledgement. I wish to thank Wolfgang May and Michael Schmidt for their insightful comments on this paper.

References

1. Baader, F., Calvanese, D., McGuinness, D.L., Nardi, D., Patel-Schneider, P.F.: The Description Logic Handbook. Cambridge University Press, Cambridge (2003)
2. Berners-Lee, T.: Relational Databases on the Semantic Web (1998), http://www.w3.org/DesignIssues/RDB-RDF.html

3. Biskup, J., Polle, T.: Adding inclusion dependencies to an object-oriented data model with uniqueness constraints. Acta Inf. 39(6-7) (2003)
4. Bizer, C.: D2R MAP - A Database to RDF Mapping Language. WWW (Posters) (2003)
5. Blakeley, C.: Mapping Relational Data to RDF with Virtuoso's RDF Views. Open-Link Software (2007)
6. van Bommel, M.F., Weddell, G.E.: Reasoning about equations and functional dependencies on complex objects. IEEE ToKDE 6(3) (1994)
7. Calvanese, D., De Giacomo, G., Lembo, D., Lenzerini, M., Rosati, R.: Can OWL model football leagues?. In: OWLED 2007 (2007)
8. Chatterjee, N., Krishna, M.: Semantic Integration of Heterogeneous Databases on the Web. In: ICCTA 2007 (2007)
9. Chen, P.P.: The Entity-Relationship Model - Toward a Unified View of Data. ACM Trans. on Database Systems 1(1) (1976)
10. Fan, W., Libkin, L.: On XML integrity constraints in the presence of DTDs. J. ACM 49(3) (2002)
11. Kifer, M., Lausen, G., Wu, J.: Logical Foundations of Object-Oriented and Frame-Based Languages. J. ACM 42(4) (1995)
12. Krishna, M.: Retaining Semantics in Relational Databases by Mapping them to RDF. In: IAT Workshops 2006 (2006)
13. Lausen, G., Meier, M., Schmidt, M.: SPARQling Constraints for RDF. In: EDBT 2008 (to appear, 2008)
14. Lutz, C., Areces, C., Horrocks, I., Sattler, U.: Keys, Nominals, and Concrete Domains. J. Artif. Intell. Res. (JAIR) 23 (2005)
15. Motik, B., Horrocks, I., Sattler, U.: Bridging the Gap Between OWL and Relational Databases. In: WWW 2007 (2007)
16. Motik, B., Horrocks, I., Sattler, U.: Adding Integrity Constraints to OWL. In: OWLED 2007 (2007)
17. Staab, S., Studer, R. (eds.): Handbook on Ontologies. Springer, Heidelberg (2004)
18. OWL Web Ontology Language Reference. W3C (2004)
19. OWL 1.1 Web Ontology Language Overview. Editor's Draft, April 6 (2007), http://webont.org/owl/1.1/overview.html
20. OWL 1.1 Web Ontology Language Model-Theoretic Semantics. Editor's Draft, April 6 (2007), http://webont.org/owl/1.1/semantics.html
21. Resource Description Framework (RDF): Concepts and Abstract Syntax. W3C (2004)
22. RDF Semantics. W3C (2004)
23. SPARQL Query Language for RDF. W3C (2007)
24. XML Schema Part 0: Primer Second Edition. W3C (2004)

An Effective SPARQL Support over Relational Databases

Jing Lu[1,2], Feng Cao[2], Li Ma[2], Yong Yu[1], and Yue Pan[2]

[1] Shanghai Jiao Tong University, Shanghai 200030, China
{robert_lu,yyu}@cs.sjtu.edu.cn
[2] IBM China Research Laboratory, Beijing 100094, China
{caofeng,malli,panyue}@cn.ibm.com

Abstract. Supporting SPARQL queries over relational databases becomes an active topic recently. However, it has not received enough consideration when SPARQL queries include restrictions on values (i.e filter expressions), whereas such a scenario is very common in real life applications. Challenges to solve this problem come from two aspects, (1) databases aspect. In order to fully utilize the well-developed SQL optimization engine, the generated SQL query is desired to be a single statement. (2) SPARQL query aspect. A practical SPARQL query often embeds several filters, which require comparisons between RDF results of different types. The type of RDF resources needs to be dynamically determined in the translation. In this paper, we propose an effective approach to support SPARQL queries over relational databases, with the above challenges in mind. To ensure the seamless translation, a novel facet-based scheme is designed to handle filter expressions. Optimization strategies are proposed to reduce the complexity of the generated SQL query. Experimental results confirm the effectiveness of our proposed techniques.

1 Introduction

The Resource Description Framework (RDF) data is often persisted in relational DBMSs by triple stores. The RDF data is represented as a collection of triples <subject, predicate, object>. SPARQL [1] is W3C's recommendation for RDF query. Based on matching graph patterns, it provides lots of facilities to extract RDF subgraphs. Given a data source D, a SPARQL query consists of a pattern which is matched against D and the values obtained from this matching.

As one of the building blocks of SPARQL query, a SPARQL filter expression restricts the graph pattern matching solutions. It provides the following functionalities: 1) The ability to restrict the value of literals and arithmetic expressions; 2) The ability to preprocess the RDF data by built-in functions. For example, isIRI(term) returns true if *term* is an IRI. Therefore, in real applications, users often submit various SPARQL queries including filter expressions to express their specific requirements on results. For example, a campus analyst may issue the following query in Figure 1, which requests IRI and optional

V. Christophides et al. (Eds.): SWDB-ODBIS 2007, LNCS 5005, pp. 57–76, 2008.
© Springer-Verlag Berlin Heidelberg 2008

```
SELECT ?person ?tel WHERE {
  ?person rdf:type bm:GraduateStudent
  {
    { ?person bm:like ?interest } UNION
    { ?person bm:love ?interest }
  } .
  OPTIONAL { ?person bm:telephone ?tel } .
  ?person bm:age ?age .
  FILTER ( ?age < 25 &&
           REGEX(STR(?interest), "Ball$") )
}
```

Fig. 1. An example of SPARQL query

telephone number of all graduated students with age less than 25 and have an interest of ball sports, in a UOBM [2] ontology (The bm:age is an extended property of the UOBM).

Previous methods on supporting SPARQL over relational databases [3,4,5] mainly supported basic SPARQL query patterns. They either ignored filter expressions due to complexity or adopted memory-based method for evaluation. For example, in Sesame [6,3] or Jena2 [7], a SPARQL query with a global filter expression is first translated into a SQL without filter expressions, and then the results are further filtered in Java program. If there are nested filter expressions in an optional query pattern, only part of the query pattern is evaluated, then the results are filtered in the program and further SQLs are issued one by one. Therefore, the evaluation of filter expressions is a time consuming operation due to the expensive I/O cost. How to efficiently and scalably process SPARQL query with filter expressions is still an open question [4,5].

A single SQL query translated from a SPARQL query can fully utilize the well-developed SQL optimization engine. This is because existing DBMS optimizers usually make a query plan based on one SQL statement. In addition, the generated single SQL query can be directly embedded into other SQL queries as a sub-query, which can be optimized by DBMSs as a whole. A seamless integration of SPARQL queries with SQL queries is attractive to real applications. However, the requirement of a single SQL statement is a strong constraint to the translation.

Mutually-interwaved filter expressions make the translation of SPARQL difficult. Filter expressions often consist of multiple functions and operators, and a filter operator may have different behaviors on different operands. When the operands are variables or complex expressions, it is usually hard to determine their actual type. Furthermore, variables may be bound to different kinds of literals in different results. That is, the type of RDF resources needs to be dynamically determined in the translation. As Figure 2, a user wants to query the students whose interest and major are the same. However, the classification information in the RDF triple data is string in some cases and the code of integer in other cases. The comparison function over string and integer literal are different

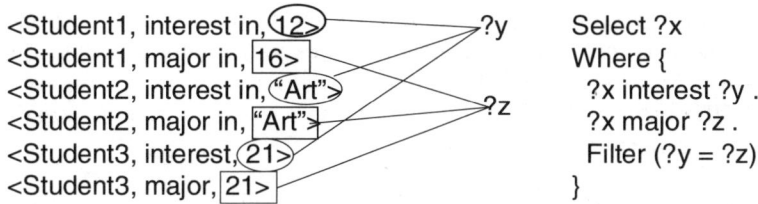

Fig. 2. An example of dynamic data type of RDF resource

in the SPARQL standard. Therefore, the generated SQL should be able to adopt different comparison methods for different types of operands.

To address the above problem, we present a complete translation process from a SPARQL query to a SQL query. The main contributions of the paper are as follows.

- We propose an effective method to translate a complete SPARQL query into a single SQL, so that the generated SQL can be directly embedded as a sub-query into other SQL queries. By that way, SPARQL queries can be seamlessly integrated with SQL queries.
- We present a novel idea of facet-based scheme to translate filter expressions into SQL statements and support most of SPARQL features, such as nested filters in optional patterns.
- We propose two optimization strategies for SQL generation, and perform experiments on benchmark data. The experimental results show the effectiveness of our method.

1.1 Preliminaries: SPARQL Pattern Tree

The generated SQL may be different for different database schemas. To simplify the discussion, we assume that all the triples are stored in a triple table, in which internal IDs are used instead of IRIs or literal strings. The IRI and literal strings are stored in two separate tables. Most known RDF stores adopt such a schema.

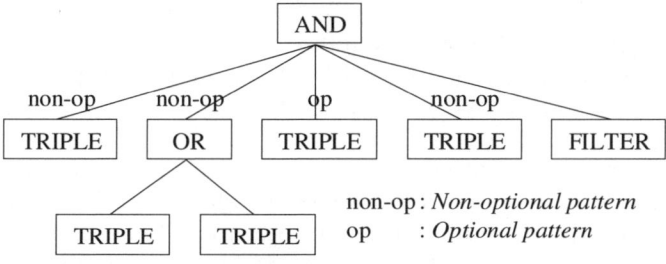

Fig. 3. An example SPARQL pattern tree

Generally, the graph pattern part in a SPARQL query can be expressed as a SPARQL pattern tree. The tree shows the backbone of the SPARQL query and is used in the translation. Similar pattern trees are used in [4]. Figure 3 shows the SPARQL pattern tree of the query in Figure 1. There are four possible types of nodes in the pattern tree:

(1) **AND node**, which corresponds to conjunction of graph patterns in a SPARQL query. In our approach, an AND node can have multiple child nodes. That is, consecutive nested AND patterns are included in one AND node as child nodes. Each child node has a flag indicating whether the node is optional or not. In this way, the optional patterns in SPARQL queries are also covered by AND nodes.

(2) **OR node**, which corresponds to a UNION pattern in a SPARQL query. Similar to the AND pattern, nested UNION patterns are represented by one OR node.

(3) **TRIPLE node**, which represents a graph pattern with a single triple. The subject, predicate and object in the triple could be constants or variables.

(4) **FILTER node**, which represents a filter expression in a SPARQL query. A FILTER node is always connected to an AND node as a special child node. In bottom-up semantics of SPARQL, only variables appearing in corresponding AND pattern can be used in the filter.

AND node, OR node and TRIPLE node represent graph patterns in the RDF graph. We call them pattern nodes.

The remainder of this paper is organized as follows. Section 2 describes the problem of filter expression translation. Section 3 presents our translation method. Section 4 proposes two optimization strategies. Section 5 presents an example of our translation. Section 6 provides experimental analysis. In section 7, we briefly survey the related work. Section 8 concludes this paper.

2   Problem Statement

A SPARQL filter can not be translated into a SQL expression straightforwardly. The result of a filter expression or a sub-expression is an RDF object, which could be a literal, an IRI reference or an error (e.g. when applying functions on unbound values). However, the result of a SQL expression is always a primitive value, such as a string, a double or an integer. Taking literals as an example, we know that a literal could have a lexical form, an optional language tag and an optional datatype. It is always hard to describe a literal object by a primitive value.

Fortunately, we found that usually only a primitive part of an RDF object is used in a function or operator. Thus, we define "facets" of RDF objects which facilitates the translation from a SPARQL filter to a SQL expression.

With the concept of facet, the problem of filter translation changes from "What is the SQL for a filter expression" to "What is the SQL for the Boolean Facet of a filter expression".

2.1   Facet of an RDF Object

A facet of an RDF object is somewhat like a view, or a data field of an object. The value of each facet is always a primitive value of a specific type. We define the following facets:

- *IRI Facet*, which is the full IRI string of an IRI reference. The primitive datatype is string. For literals, this facet is not available.
- *Lexical Facet*, which is the lexical form of a literal. The primitive datatype is string. This facet is only available for literals.
- *Language Facet*, which is the language tag of a literal. The primitive datatype is string. This facet is only available for literals. If the literal is a typed literal or a plain literal without a language tag, the value is an empty string. This is consistent with the definition of the built-in SPARQL function "Lang" .
- *Datatype Facet*, which is the full IRI of the datatype of a typed literal. The primitive datatype is string. This facet is only available for literals. For plain literals, the value is a NULL.
- *Numeric Facet*, which is the numeric value of a numeric literal. The primitive datatype is double. This facet is only available for typed literals with datatype xsd:float, xsd:double, xsd:decimal or a sub-type of xsd: decimal.
- *Boolean Facet*, which is the boolean value of a boolean literal. Boolean Facet is translated into SQL predicates, such as "a=b" or "a IS NULL". This facet is only available for typed literals with datatype xsd:boolean.
- *Date time Facet*, which is the date time value of a date time literal. In implementation, we map a date time value into a 64-bit integer, so that the nature time order keeps in this mapping. Thus, the primitive datatype is 64-bit integer. This facet is only available for typed literals with datatype xsd:dateTime.
- *ID Facet*, which is the internal ID of the RDF object. The primitive datatype is integer. As a constant or a calculated result is not required to have a corresponding internal ID, this facet is only available for variables.

IRI, Lexical, Language and Datatype Facets express natural attributes of RDF objects. Numeric, Boolean and Date time Facets express typed literals in corresponding native data types of databases, which facilitates the translation of filter expressions. ID Facet is used to check whether a variable is bound or not.

3   SPARQL to SQL Translation

In this section, we discuss in detail our facet-based SPARQL to SQL translation approach. We refer to it as the FSparql2Sql. The FSparql2Sql consists of two parts: translation of pattern nodes and translation of filter expressions.

3.1   Translation of Pattern Nodes

Similar to the approach in [4], each pattern node is translated into a SQL sub-query. For each variable in the pattern, there is a corresponding column in the query result, which contains the ID of the candidate result. If a variable is unbound, a NULL is returned in that column instead.

- A TRIPLE node is translated into a simple SELECT query on the triple table. If there are constants in the triple pattern, a corresponding WHERE clause is added to the query. Otherwise, the columns are renamed to the corresponding variable names. When one variable appears multiple times in the triple pattern, an equivalent constraint of these columns should be added to the WHERE clause. For example, a triple pattern:

```
{?person bm:isFriendOf ?person}
```

is translated into:

```
SELECT subject AS person FROM triple
    WHERE predicate = pID and subject = object
```

Here, the pID stands for the ID the IRI <bm:isFriendOf>.
- An AND node is translated into a query on consequent joins of the sub-queries from its child pattern nodes.

  If a child node is optional, a left join is used instead of an inner join. When all the child nodes are optional, a dummy table should be appended as the first table in the joined table list, so that other sub queries can be left joined to this table. The dummy table is the table containing only one line and one column.

  In addition, we keep tracing each variable on whether it could be possibly unbound. The different states will affect the join conditions used. Also, the COALESCE function might be used to combine results from multiple queries.

  Finally, if there are child FILTER nodes, they are translated into SQL expressions and added to the WHERE clause of the query.
- An OR node is translated into a UNION of the sub-queries. If the variable sets of the sub-queries do not match each other, dummy columns containing constant NULL should be added to the sub-queries.

3.2 Translation of Filter Expressions

We support the primary operators and functions of filter expressions, which includes:

- (1) IRI constants and literal constants;
- (2) named variables;
- (3) calculation operators (+, -, *, /);
- (4) comparison operators (=, !=, >, <, >=, <=);
- (5) logical operators (&&, ||, !);
- (6) built-in functions (bound, isIRI, isLiteral, datatype, lang, str, regex).

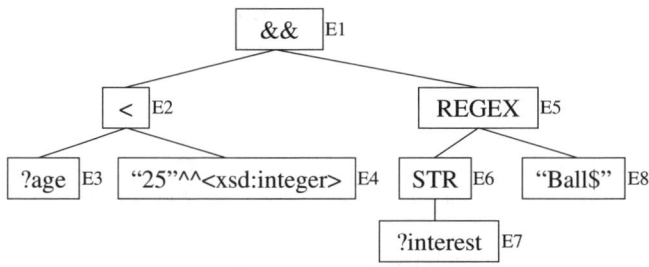

Fig. 4. A filter expression tree

A SPARQL filter expression can be parsed into a filter expression tree. Non-leaf nodes are operators or functions, while leaf nodes are constants or variables. Figure 4 shows the filter expression tree of the `FILTER` clause in Figure 1.

By a filter expression tree, the goal of the translation is to get the SQL representing the Boolean Facet of the root node. For a specific facet of an expression node, its translation result is recursively constructed from generating SQLs of its child nodes.

In the following, we adopt "`X.Lexical`" to represent the generated SQL for the Lexical Facet of "X". Similarly, "`X.ID`" represents the SQL for the ID Facet of "X", and so on.

Literal Constants and IRI Constants. The translation of constants in a filter expression is rather straightforward. The translation results for all kinds of supported facets are just SQL constants as described in the facet definitions. For literal constants, the lexical, language and datatype facet is always available. For typed literals, the numeric, boolean and date time facet might also be available. For IRI reference constants, the only available facet is the IRI facet.

Variables. As internal IDs are used to encode IRIs and literals, the IRI table and literal table should be used to get the actual IRI or literal value of the variable.

In order to effectively translate the Numeric, Boolean and Date Time Facets of variables, we suggest to three additional tables in the database schema, which map the literal IDs to these facet values, respectively.

Therefore, if the SQL for a specific facet of a variable is requested, the corresponding table is added (using left join) to the SQL query, and the corresponding column in that table is provided as the translation result.

In order to avoid unnecessary join operations, during the translation of graph pattern nodes, we keep tracking the possible binding value of each variable. For example, if a variable appears in the subject or predicate position of a triple, the value of the variable cannot be a literal. Thus, if the SQL for a literal related facet is requested, we can directly return a `NULL`.

Comparison Operators. The translation of comparison operators is the most complex in all operators. Because the non-equality operators (`>`, `<`, `>=` and `<=`)

can be used between numeric literals, simple literals, string literals, boolean literals or date time literals, while the equality operators (= and !=) can be further used between all kinds of RDF terms. According to SPARQL specifications, different comparison methods should be used for different types of operands.

CASE Expression. It is often hard to determine the types of operands, especially when the operands are complex expressions. The types of operands need to be bound dynamically. To deal with this problem, we choose to use SQL *CASE Expression* in the translation. A *CASE Expression* often consists of several WHEN clauses, in order to provide different results for different cases. The CASE...WHEN statement is defined in ANSI SQL-92 standard which is widely supported by many existing DBMSs.

We use the Lexical Facet of operator ">" as an example. For each facet available in both operands, we add a corresponding WHEN clause to the CASE expression.

1) If the Numeric Facet is available, add

```
WHEN X.Numeric > Y.Numeric THEN 'true'
WHEN X.Numeric <= Y.Numeric THEN 'false'
```

2) If the DateTime Facet is available, add

```
WHEN X.DateTime > Y.DateTime THEN 'true'
WHEN X.DateTime <= Y.DateTime THEN 'false'
```

3) If the Boolean Facet is available, add

```
WHEN X.Boolean AND NOT Y.Boolean THEN 'true'
WHEN NOT X.Boolean OR Y.Boolean THEN 'false'
```

4) If the Lexical Facet is available, add

```
WHEN X.Language = '' AND Y.Language = '' THEN
  CASE WHEN X.Datatype IS NULL
       AND Y.Datatype IS NULL THEN
    CASE WHEN X.Lexical > Y.Lexical THEN 'true
    WHEN X.Lexical <= Y.Lexical THEN 'false' END
  WHEN X.Datatype = xsd:string
    AND Y.Datatype = xsd:string THEN
    CASE WHEN X.Lexical > Y.Lexical THEN 'true'
    WHEN X.Lexical <= Y.Lexical THEN 'false' END
  END
```

* The xsd:string stands for its full URI SQL string.

After the above steps, we obtain the final translation result, which is probably a complex CASE expression. We will discuss how to optimize it in Section 4.

Built-in Functions. We can see that all these built-in functions always return an IRI or a literal with a fixed type as the result (except for error results). For example, the `bound`, `isIRI`, `isLiteral` and `regex` functions always return a boolean typed literal. The `lang` and `str` functions always return a plain literal with no language tag. The `datatype` function always returns an IRI reference.

The translation for Language and Datatype Facet is simply a string constant decided by the function or operator. For other facets, when error happens, it is ensured that a NULL is returned instead of the normal results. In order to reduce the complexity of the translation, a constant is always returned, no matter error happens or not. Thus, only the lexical(or boolean) facet is used in the translation to determine if an expression gives an error result or not. We take `bound` functions as an example below.

The `bound` functions always return a boolean typed literal. As described above, the language facet and datatype facet will always return a constant. The boolean facet is translated to an "IS NULL" predicate on the ID facet, IRI facet or lexical facet of the operand expression, respectively. In some rare cases, the lexical facet of these functions is required. The result of the lexical facet should be the canonical representation of the boolean values, that is, string "`true`" or "`false`". In this case we have to use a CASE expression over the "IS NULL" predicate to give the string result.

Calculation Operators. The calculation operators include +, -, * and /. These operators may only be used between numeric literals and always gives a numeric literal as the result. As described above, we always use double as the datatype in our translation for numeric values. Thus, for the numeric facet, the translation result is simply the calculation operator over the numeric facet of the operands. When the lexical facet is required, we have to use a SQL function to convert the numeric result into a string. Note that the numeric-to-string conversion function is not contained in SQL standard and thus may various between different databases.

Logical Operators. The logical operators include \&\&, || and !. These operators may only be used between boolean literals and always returns a boolean literal as the result. The language and datatype facet will always give a constant as the result. For the boolean facet, the translation result is simply the corresponding SQL logical operator (AND, OR or NOT) over the boolean facet of the operands. If the lexical facet is required, we have to use an additional CASE expression to change the result of the SQL predicate into a string "`true`" or "`false`". The SQL will look like this. (`E.boolean` represents the translate result of the boolean facet of the whole expression.)

```
CASE WHEN E.boolean THEN 'true'
     WHEN NOT E.boolean THEN 'false'
END
```

Here, a second WHEN clause is used instead of the ELSE clause because when the `E.boolean` returns an UNKNOWN as the result, the result of the lexical facet should be an NULL instead of the "false". Furthermore, the SPARQL

Table 1. logical-AND and logical-OR truth table

A	B	A || B	A && B	A	B	A || B	A && B
T	T	T	T	E	T	T	E
T	F	T	F	F	E	E	F
F	T	T	F	E	F	E	F
F	F	F	F	E	E	E	E
T	E	T	E				

Table 2. Translation of build-in functions

Function	Generated SQLs For Different Facets				
	Boolean	Lexical	Lang.	Datatype	IRI
bound(X)	X.ID IS NOT NULL	CASE WHEN X.ID IS NOT NULL THEN 'true' ELSE 'false' END	An empty string.	The IRI of xsd:boolean	(N/A)
isIRI(X)	X.IRI IS NOT NULL	CASE WHEN X.IRI IS NOT NULL THEN 'true' ELSE 'false' END	An empty string.	The IRI of xsd:boolean	(N/A)
isLiteral(X)	X.Lexical IS NOT NULL	CASE WHEN X.Lexical IS NOT NULL THEN 'true' ELSE 'false' END	An empty string.	The IRI of xsd:boolean	(N/A)
datatype(X)	(N/A)	(N/A)	(N/A)	(N/A)	X.Datatype
lang(X)	(N/A)	X.Language	An empty string.	A NULL constant.	(N/A)
str(X)	(N/A)	COALESCE(X.IRI, X.Lexical)	An empty string.	A NULL constant.	(N/A)
regex(X, pattern)	X.Lexical LIKE likePattern (limited by the expressiveness of LIKE operator)	CASE WHEN X.Lexical LIKE likePattern THEN 'true' WHEN X.Lexical NOT LIKE likePattern THEN 'false' END	An empty string.	The IRI of xsd:boolean	(N/A)

* X.Lexical represents the generated SQL for the Lexical Facet of "X";
* X.ID represents the SQL for the ID Facet of "X", and so on.

logical-AND and logical-OR truth table for true (T), false (F), and error (E) [1] is as following Table 1, which just corresponds with the semantic of SQL logical-AND and logical-OR.

Tables 2 and 3 show the translations of built-in functions, calculation operators and logical operators on various facets.

Table 3. Translation of Calculation Operators and Logical Operators

Ops	Generated SQLs For Different Facets				
	Numeric	Boolean	Lexical	Lang.	Datatype
X *op* Y (*op* is +, −, * or /)	X.Numeric *op* Y.Numeric	(N/A)	CHAR(X.Numeric *op* Y.Numeric) The number-to-string conversion function may be different in various databases.	An empty string.	The IRI of xsd:double.
X *op* Y (*op* is && or ||)	(N/A)	X.Boolean AND Y.Boolean for && X.Boolean OR Y.Boolean for ||	CASE WHEN X.Boolean AND Y.Boolean THEN 'true' WHEN NOT (X.Boolean AND Y.Boolean) THEN 'false' END For ||, replace the "AND" with "OR" in the SQL.	An empty string.	The IRI of xsd:boolean.
! X	(N/A)	NOT X.Boolean	CASE WHEN X.Boolean THEN 'false' WHEN NOT X.Boolean THEN 'true' END	An empty string.	The IRI of xsd:boolean.

* X.Lexical represents the generated SQL for the Lexical Facet of "X";
* X.ID represents the SQL for the ID Facet of "X", and so on.

4 Optimization

Our facet-based translation may generate very complex result SQL statement. For example, in order to meet the requirement of dynamic types of operands, CASE expressions and lots of constants will appear in the generated SQL. As far as we know, most database optimizers (e.g., DB2) can not perform good optimization over complex constant and CASE expressions, since these expressions are not commonly appeared in hand-written SQLs. Therefore, additional optimization over CASE expression and constant are needed.

4.1 Optimization on CASE Expression

The CASE expression in generated SQLs can sometimes be replaced by other expressions which can be well optimized by DBMS engines. For example, by our algorithm, a simple filter expression "?X != <http://foo/boo>" is translated into the following SQL expression. The t1 is a local nickname of the IRI table.

```
(CASE WHEN t1.IRI <> 'http://foo/boo' THEN 'true'
      WHEN t1.IRI = 'http://foo/boo' THEN 'false'
 END) = 'true'
```

However, the above SQL expression can be simply rewritten into another SQL expression with the same semantics:

```
t1.IRI <> 'http://foo/boo'
```

In order to solve this problem, we apply a special optimization on CASE expressions. We try to find such "=" predicates in the expression which meet all following requirements:

1. One side is a constant while the other side is a CASE expression with exactly two WHEN clauses.
2. Predicates of the two WHEN clauses are exactly the negation of each other. There are two kinds of negations. One is replacing the main operator with the inverse one, while the other is adding a NOT outside.
3. One of the results matches the constant on the other side of the "=" while the other does not.

If such a predicate is found, we replace the whole predicate with the predicate inside the WHEN clause, whose result matches the constant.

4.2   Optimization on Constant Expression

Optimization on constant expressions are especially important for the auto-generated SQLs. One reason is that lots of constants (such as results of Language Facet, Datatype Facet, etc.) may appear in the SQL expression, while most existing database optimizers can not perform good optimization over complex constant expressions.

Another reason is that after this process, it is more likely that the optimization on CASE expression can be applied, as useless WHEN clauses might be totally removed in the process. Therefore, we recursively optimize the constant expressions in the generated SQL. Table 4 shows several examples. <A> and <B> represent arbitrary sub-expressions.

Table 4. Constant expression optimization examples

Before Optimization	After Optimization
'ab' <> ''	0=0
'ab' IS NULL	0<>0
<A> AND 0=0	<A>
<A> OR 0=0	0=0
COALESCE(<A>, 'ab', <B>)	COALESCE(<A>, 'ab')
COALESCE(<A>, NULLIF('',''), <B>)	COALESCE(<A>, <B>)
CASE WHEN 0<>0 THEN <A> WHEN <B> ...	CASE WHEN <B> ...
CASE WHEN 0=0 THEN <A> WHEN <B> ...	<A>

5   An Example of Translation

In this section, we take the query in Figure 1 as an example to explain the translation for filter expressions. The E1 to E8 in Figure 4 are the names of the expression nodes.

(1) The goal of translation is to generate SQLs for the Boolean Facet of the root node E1. E1 is an expression node with && operator. By the translation of && in Table 3, we get

```
E1.Boolean = E2.Boolean AND E5.Boolean
```

(2) E2 is an expression node with < operator. By the processes in Section 3.2, we get the (2.1) to (2.3) steps.

(2.1) As E3 and E4 can provide a Numeric Facet, the SQLs for E2.Boolean includes the following WHEN clauses:

```
WHEN E3.Numeric < E4.Numeric THEN 'true'
WHEN E3.Numeric >= E4.Numeric THEN 'false'
```

(2.2) As E3 and E4 can provide a Lexical Facet, the SQLs for E2.Boolean further includes the following WHEN clause:

```
WHEN E3.Language = ' ' AND E4.Language = ' ' THEN
  CASE WHEN E3.Datatype IS NULL
       AND E4.Datatype IS NULL THEN
    CASE WHEN E3.Lexical < E4.Lexical THEN 'true'
         WHEN E3.Lexical >= E4.Lexical THEN 'false'
  WHEN E3.Datatype = xsd:string
   AND E4.Datatype = xsd:string THEN
    CASE WHEN E3.Lexical < E4.Lexical THEN 'true'
         WHEN E3.Lexical >= E4.Lexical THEN 'false'
  END
```

*   The xsd:string stands for its full URI SQL string.

(2.2.1) As E3 ?age is a variable node, as the description in Section 3.2, two left joins (one for Numeric Facet, the other for Lexical, Language and Datatype Facets) are added to the SQLs of the parent AND node, and the corresponding columns are returned as the result of these facets.

(2.2.2) E4 is a literal constant node, so the results of the requested facets are as follows:

```
E4.Numeric = 25.0   E4.Lexical = '25'
E4.Language = ' '      E4.Datatype = xsd:integer
```

(2.3) Adding a "= 'true'" after the big CASE expression, we get the complete translation result of E2.Boolean.

(3) E5 is a REGEX function, by Table 2 the Boolean Facet is translated to "E6.Lexical LIKE '%Ball'". The '%Ball' is translated from the regular expression pattern "Ball$".

(3.1) The Lexical Facet of E6 should be translated to the IRI Facet or Lexical Facet of E7. However, we can know from the triple patterns that variable ?interest is bound to objects of object property "bm:like" or "bm:love", and thus must be IRI references. So the translation of E6.Lexical becomes simply E7.IRI. In order to translate E7.IRI, one more left join is added in order to retrieve the actual IRI string for variable ?interest.

(4) Now we have the following translation result before the optimization step:

```
(CASE
   WHEN ?age.Numeric < 25.0 THEN 'true'
   WHEN ?age.Numeric >= 25.0 THEN 'false'
   WHEN ?age.Language = ' ' AND ' ' = ' ' THEN
      CASE WHEN ?age.Datatype IS NULL
            AND xsd:integer IS NULL THEN
         CASE WHEN ?age.Lexical < '25' THEN 'true'
               WHEN ?age.Lexical >= '25' THEN 'false'
      WHEN ?age.Datatype = xsd:string
        AND xsd:integer = xsd:string THEN
         CASE WHEN ?age.Lexical < '25' THEN 'true'
               WHEN ?age.Lexical >= '25' THEN 'false'
      END
END) = 'true' AND ?interest.IRI LIKE '%Ball'
```

Notice that ?age.Numeric corresponds to the value column in Numeric table, and so forth. The SQL result includes several CASE clauses for the comparisons of ?age with different possible data types.

Optimization for the generated SQL. The optimization includes following steps:

(1) By the the optimizations on constant expressions, "xsd:integer IS NULL" and "xsd:integer = xsd:string" can be replaced with false constants, i.e., "0<>0".

(2) The two surrounding WHEN clauses can be totally removed as the condition is never true. Now, the nested CASE expression have no more WHEN clauses inside and is replaced by "NULLIF(' ',' ')".

(3) The clause "WHEN ?age.Language = ' ' AND ' ' = ' ' THEN NULLIF(' ',' ')" can be removed, as it is the last WHEN clause and the result is the same as the ELSE part, which is NULL by default. The SQL becomes:

```
(CASE
   WHEN ?age.Numeric < 25.0 THEN 'true'
   WHEN ?age.Numeric >= 25.0 THEN 'false'
END) = 'true' AND ?interest.IRI LIKE '%Ball'
```

(4) It can be seen that the "=" predicate meets the requirements for the CASE expression optimization described in Section 4.1. We get the final translate result after this optimization, which is a very simple SQL:

```
?age.Numeric < 25.0 AND ?interest.IRI LIKE '%Ball'
```

6   Experimental Analysis

All experiments were run on a 3.0GHz Core 2 Duo PC with 2GB RAM, running Microsoft Windows XP Professional. IBM DB2 9.1 Enterprise edition is used as the backend store. We implemented the FSparql2Sql in SOR [8]. The Sesame v1.2.6 [3] is extended with DB2 support and adopted for comparison purpose.

To better evaluate filter expressions, we extend the University Ontology Benchmark (UOBM)[2] with a new property "bm:age". Every student and professor are given an integer age. The largest dataset includes 1.1M instances and 2.4M relationship assertions. Six adopted SPARQL queries are shown in Table 5. We further adopted Query 7 to Query 25 for testing the functionality of FSparql2Sql on built-in functions (refer to Table 6), comparison operators (refer to Table 7), logical operators (refer to Table 8) and calculation operators (refer to Table 9), respectively.

Effectiveness Analysis. As shown in Figure 5(a), the execution time of Sesame with a global filter (i.e., Q1) or not (i.e., Q0) is almost the same. Because Sesame uses the same SQL for them and evaluates global filters in Java program. The queries with a nested filter (i.e., Q2) is definitely slower than the ones without it. The reason is that lots of SQL queries are generated and executed for the nested filter. Figure 6(a) show the same tendency for a more complex group of queries.

For FSparql2Sql, we can observe from Figure 6(b) that the queries with a global filter run faster than the one without it, while the queries with a nested

Table 5. SPARQL Queries

Groups	Queries	Notes

Group 1	Q0. SELECT ?x ?y WHERE {
	    ?x rdf:type bm:GraduateStudent .
	    ?x bm:age ?y
	}
		Q0 is a simple triple pattern query without Filter.

	Q1. SELECT ?x ?y WHERE {
	    ?x rdf:type bm:GraduateStudent .
	    ?x bm:age ?y .
	    FILTER (?y < 25)
	}
		Q1 extends Q0 with global filters.

	Q2. SELECT ?x ?y WHERE {
	    ?x rdf:type bm:GraduateStudent .
	    OPTIONAL {
	        ?x bm:age ?y .
	        FILTER (?y < 25)
	    }
	}
		Q2 extends Q0 with nested filters in optional patterns.

Group 2	Q3. SELECT ?x ?y WHERE {
	    ?x rdf:type bm:GraduateStudent .
	    ?x bm:isAdvisedBy ?y .
	    ?y bm:age ?z
	}
		Q3 is a complex triple pattern query without Filter.

	Q4. SELECT ?x ?y WHERE {
	    ?x rdf:type bm:GraduateStudent .
	    ?x bm:isAdvisedBy ?y .
	    ?y bm:age ?z .
	    FILTER (?z > 50)
	}
		Q4 extends Q3 with global filters.

	Q5. SELECT ?x ?y WHERE {
	    ?x rdf:type bm:GraduateStudent .
	    OPTIONAL {
	        ?x bm:isAdvisedBy ?y .
	        ?y bm:age ?z .
	        FILTER (?z > 50)
	    }
	}
		Q5 extends Q3 with nested filters in optional patterns.

Table 6. SPARQL Queries of built-in functions

Groups	Queries	Notes

Group3 (built-in functions)	Q7. SELECT ?x ?y WHERE {
	    ?x rdf:type bm:GraduateStudent .
	    ?x bm:age ?y .
	    FILTER (bound(?x))
	}
		function bound

	Q8. SELECT ?x ?y WHERE {
	    ?x rdf:type bm:GraduateStudent .
	    ?x bm:age ?y .
	    FILTER (isIRI(?x))
	}
		function isIRI

	Q9. SELECT ?x ?y WHERE {
	    ?x rdf:type bm:GraduateStudent .
	    ?x bm:age ?y .
	    FILTER (isLiteral(?y))
	}
		function isLiteral

	Q10.SELECT ?x ?y WHERE {
	    ?x rdf:type bm:GraduateStudent .
	    ?x bm:age ?y .
	    FILTER (datatype(?y) = xsd:integer)
	}
		function datatype

	Q11.SELECT ?x WHERE {
	    ?x rdf:type bm:GraduateStudent .
	    ?x bm:like ?interest .
	    FILTER (lang(?interest) = EN") "
	}
		function lang

	Q12.SELECT ?x WHERE {
	    ?x rdf:type bm:GraduateStudent .
	    ?x bm:like ?interest .
	    FILTER (REGEX(STR(?interest), Ball$"))"
	}
		function STR

Table 7. SPARQL Queries of Comparison Operators

Groups	Queries	Notes

Group4 (comparison operators)	Q13.SELECT ?x ?y WHERE {
	    ?x rdf:type bm:GraduateStudent .
	    ?x bm:age ?y .
	    FILTER (?y < 25)
	}
		operator <

	Q14.SELECT ?x ?y WHERE {
	    ?x rdf:type bm:GraduateStudent .
	    ?x bm:age ?y .
	    FILTER (?y <= 25)
	}
		operator <=

	Q15.SELECT ?x ?y WHERE {
	    ?x rdf:type bm:GraduateStudent .
	    ?x bm:age ?y .
	    FILTER (?y > 25)
	}
		operator >

	Q16.SELECT ?x ?y WHERE {
	    ?x rdf:type bm:GraduateStudent .
	    ?x bm:age ?y .
	    FILTER (?y >= 25)
	}
		operator >=

	Q17.SELECT ?x ?y WHERE {
	    ?x rdf:type bm:GraduateStudent .
	    ?x bm:age ?y .
	    FILTER (?y = 25)
	}
		operator =

	Q18.SELECT ?x ?y WHERE {
	    ?x rdf:type bm:GraduateStudent .
	    ?x bm:age ?y .
	    FILTER (?y != 25)
	}
		operator !=

filter use almost the same time as the normal ones. Figure 6(b) show the same tendency for a more complex group of queries. This is because FSparql2Sql successfully translates the global filter directly into a WHERE condition, and translates the nested filter expressions into left joins with WHERE conditions.

Comparing Figures 5(a) and (b), we can see that the FSparql2Sql is almost 50 times faster than Sesame even for queries without filters. This is mainly due to the differences between two systems, such as different database schemas and the batch strategy adopted in FSparql2SqlAnd FSparql2Sql clearly outperforms

Table 8. SPARQL Queries of Logical Operators

Groups	Queries	Notes

Group6 (logical operators)	Q19.SELECT ?x ?y WHERE {
	    ?x rdf:type bm:GraduateStudent .
	    ?x bm:age ?y .
	    FILTER (isLiteral(?y) && isURI(?x))
	    }
		operator &&

	Q20.SELECT ?x ?y WHERE {
	    ?x rdf:type bm:GraduateStudent .
	    ?x bm:age ?y .
	    FILTER (isLiteral(?y) || isURI(?x))
	    }
		operator ||

	Q21.SELECT ?x WHERE {
	    ?x rdf:type bm:GraduateStudent .
	    ?x bm:like ?interest .
	    FILTER (! REGEX(STR1(?interest), Ball$"))"
	    }
		operator !

Table 9. SPARQL Queries of Calculation Operators

Groups	Queries	Notes

Group7 (calculation operators)	Q22.SELECT ?x ?y WHERE {
	    ?x rdf:type bm:GraduateStudent .
	    ?y rdf:type bm:GraduateStudent .
	    ?x bm:age ?age1 .
	    ?y bm:age ?age2 .
	    FILTER ((?age1 + ?age2) > 40)
	    }
		operator +

	Q23.SELECT ?x ?y WHERE {
	    ?x rdf:type bm:GraduateStudent .
	    ?y rdf:type bm:GraduateStudent .
	    ?x bm:age ?age1 .
	    ?y bm:age ?age2 .
	    FILTER ((?age1 - ?age2) > 0)
	    }
		operator -

	Q24.SELECT ?x ?y WHERE {
	    ?x rdf:type bm:GraduateStudent .
	    ?y rdf:type bm:GraduateStudent .
	    ?x bm:age ?age1 .
	    ?y bm:age ?age2 .
	    FILTER ((?age1 / ?age2) = 1)
	    }
		operator /

	Q25.SELECT ?x ?y WHERE {
	    ?x rdf:type bm:GraduateStudent .
	    ?y rdf:type bm:GradudateStudent .
	    ?x bm:age ?age1 .
	    ?y bm:age ?age2 .
	    FILTER ((?age1 * ?age2) > 400)
	    }
		operator *

Sesame about 100 and 150 times for global and nested filter, respectively. This is because FSparql2Sql fully utilizes the filter information to generate more efficient SQLs.

Scalability and Functionality Test. To test the scalability of FSparql2Sql we adopt a simple query of Query 1 and a more complex query as shown in Figure 1 (we numbered it as Query 6). Figure 7(b) shows the execution time of FSparql2Sql increase slowly with the number of universities. Figure 7(a) shows the results of Sesame over the same queries. Compared Figures 7(a) and (b), we can see that FSparql2Sql achieves average 100 performance gain over Sesame in Queries 1 and 6.

FSparql2Sql supports almost all filter expressions in SPARQL. In order to show the functionality of FSparql2Sql we test the queries defined in Tables 6, 7, 8 and 9 over thress universities. Figure 8, 9, 10 and 11 show our FSparql2Sql can

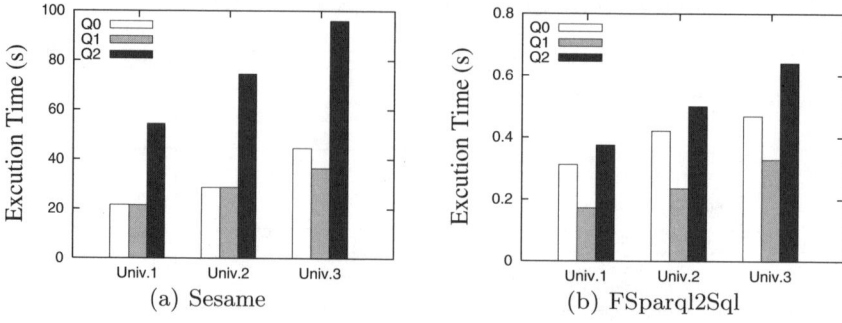

Fig. 5. Execution Time on Query Group 1

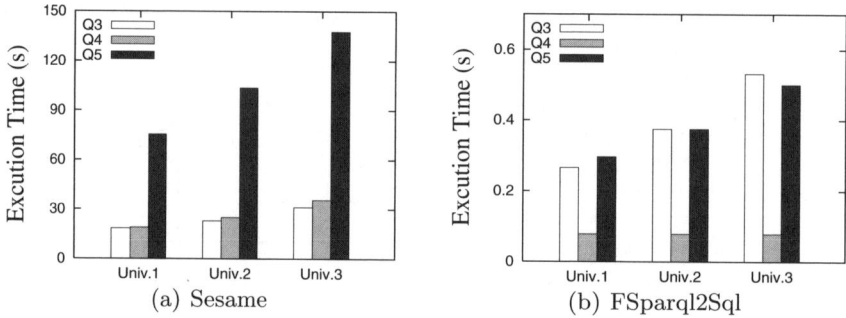

Fig. 6. Execution Time on Query Group 2

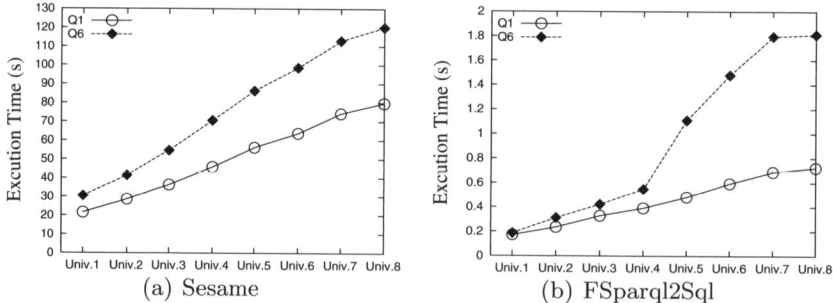

Fig. 7. Execution Time on Queries 1 and 6

well support all these built-in functions and operators and almost show linearly increase over the number of universities.

Therefore, we claim that our FSparql2Sql is an effective method to generate more efficient SQLs and better utilize the DBMS optimizers.

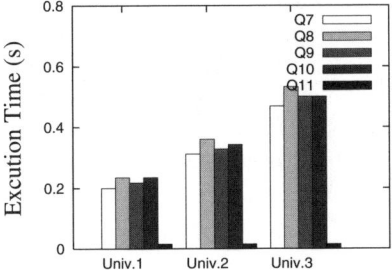

Fig. 8. Built-in Functions

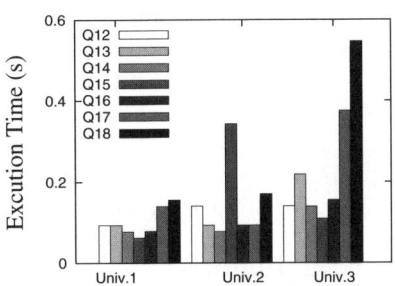

Fig. 9. Comparison Operators

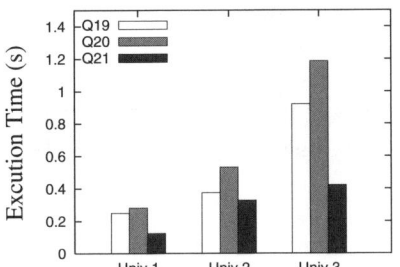

Fig. 10. Logical Operators

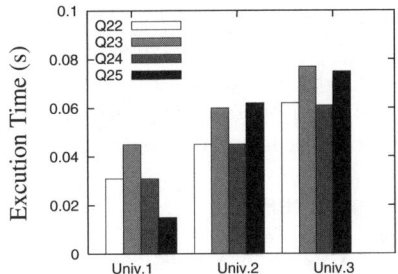

Fig. 11. Calculation Operators

7   Related Work

There are research work around semantics of SPARQL [9,1]. Jorge Perez et al. [9] provided a compositional semantics and complexity bounds, and showed that the evaluation of SPARQL patterns is PSPACE-complete without filter conditions. Different from these works on SPARQL itself, our method mainly focuses on supporting SPARQL queries over relational DBs.

Some other works proposed methods to support basic query patterns of SPARQL [3,4,5,10], including triple patterns, union patterns and optional patterns. For filter expression, They either ignore it due to complexity, or evaluate it based on memory. Cyganiak [4] presented a relational model of SPARQL and defined relational algebra operators (join, left outer join, projection, selection, etc.) to model SPARQL. Using relational algebra operators similar to [4], Harris [5] presented an implementation of SPARQL queries on top of a relational database engine, but only including a subset of SPARQL (there are no UNION operator and nested optional blocks). Sparql2Sql [7] and Sesame [6] query module are two famous query engines for SPARQL. They rewrite SPARQL queries into SQL statements. However, filter expressions are not handled by the database, but evaluated in Java code. In this paper, we show effective schemes to translate practical filter expressions into SQL statements so that a SPARQL query can be completely represented by a SQL query.

A SQL table function is defined in Oracel 10gR2 to accept RDF queries as a subquery [11], instead of supporting new query languages, e.g. RDQL[12]. RDF graph patterns defined in the table function as input are similar to the basic SPARQL patterns and can be considered as a small subset of SPARQL in terms of query capability. Compared with Oracle's RDF query based on triple patterns, we support most SPARQL features, such as nested optional patterns and complex filter expressions.

8   Conclusions

Aiming to a seamless integration of SPARQL queries with SQL queries, we proposed an effective method to translate a complete SPARQL query into a single SQL. The translated SQL query can be directly used as a sub-query by other SQL queries and evaluated by well-optimized relational database engine. In particular, we proposed effective schemes to translate filter expressions into SQL statements, which is ignored or not well addressed by existing methods. Finally, we investigated optimization strategies to improve query performance significantly. Future work is to support more SPARQL features, such as XQuery functions.

References

1. SPARQL Query Language for RDF,
   http://www.w3.org/TR/2008/REC-rdf-sparql-query-20080115/
2. Ma, L., Yang, Y., Qiu, Z., Xie, G., Pan, Y., Liu, S.: Towards a Complete OWL Ontology Benchmark. In: Sure, Y., Domingue, J. (eds.) ESWC 2006. LNCS, vol. 4011, pp. 125–139. Springer, Heidelberg (2006)
3. Broekstra, J., Kampman, A., Harmelen, F.: Sesame: A Generic Architecture for Storing and Querying RDF and RDF Schema. In: Horrocks, I., Hendler, J. (eds.) ISWC 2002. LNCS, vol. 2342, pp. 54–68. Springer, Heidelberg (2002)
4. Cyganiak, R.: A Relational Algebra for SPARQL. HP-Labs Technical Report, HPL-2005-170 (2005)
5. Harris, S., Shadbolt, N.: SPARQL Query Processing with Conventional Relational Database Systems. In: Dean, M., Guo, Y., Jun, W., Kaschek, R., Krishnaswamy, S., Pan, Z., Sheng, Q.Z. (eds.) WISE-WS 2005. LNCS, vol. 3807, pp. 235–244. Springer, Heidelberg (2005)
6. Sesame, http://www.openrdf.org/
7. Sparql2Sql, http://jena.sourceforge.net/sparql2sql/
8. Lu, J., Ma, L., Zhang, L., Wang, C., Brunner, J., Yu, Y., Pan, Y.: SOR: A Practical System for Ontology Storage, Reasoning and Search. In: 33rd International Conference on Very Large Data Bases, pp. 1402–1405. ACM, New York (2007)
9. Pérez, J., Arenas, M., Gutierrez, C.: Semantics and Complexity of SPARQL. In: Cruz, I., Decker, S., Allemang, D., Preist, C., Schwabe, D., Mika, P., Uschold, M., Aroyo, L.M. (eds.) ISWC 2006. LNCS, vol. 4273, pp. 30–43. Springer, Heidelberg (2006)
10. Pan, Z., Heflin, J.: DLDB: Extending Relational Databases to Support Semantic Web Queries. In: 1st International Workshop on Practical and Scalable Semantic Systems. CEUR-WS.org, Florida (2003)
11. Chong, E., Das, S., Eadon, G., Srinivasan, J.: An Efficient SQL-based RDF Querying Scheme. In: 31st International Conference on Very Large Data Bases, pp. 1216–1227. ACM, New York (2005)
12. RDQL - A Query Language for RDF, http://www.w3.org/Submission/RDQL/

Implementing the COntext INterchange (COIN) Approach through Use of Semantic Web Tools

Mihai Lupu[1] and Stuart Madnick[2]

[1] Singapore-MIT Alliance, National University of Singapore
mihailup@comp.nus.edu.sg
[2] Sloan School of Management, Massachusetts Institute of Technology
smadnick@mit.edu

Abstract. The COntext INterchange (COIN) strategy is an approach to solving the problem of interoperability of semantically heterogeneous data sources through context mediation. The existing implementation of COIN uses its own notation and syntax for representing ontologies. More recently, the OWL Web Ontology Language is becoming established as the W3C recommended ontology language. A bridge is needed between these two areas and an explanation on how each of the two approaches can learn from each other. We propose the use of the COIN strategy to solve context disparity and ontology interoperability problems in the emerging Semantic Web both at the ontology level and at the data level. In this work we showcase how the problems that arise from context-dependant representation of facts can be mitigated by Semantic Web techniques, as tools of the conceptual framework developed over 15 years of COIN research.

1 Introduction

Making computers understand humans is, generously put, a hard task. One of the main reasons for which this is such a hard task is because even humans cannot understand each other all the time. Even if we all spoke the same language, there still exist plenty of opportunities for misunderstanding. An excellent example is that of measure units. Again, we don't even have to go across different names to find differences: in the US, a *gallon* (the so-called Winchester gallon) is approximately 3785 ml while in the UK, the "same" *gallon* is 4546 ml, almost 1 liter more. So when we find a piece of information in a database on cars, for instance, and we learn that a particular model has a fuel tank capacity of 15 gallons, how much gas can we actually fit inside, and, consequently, for how long can we drive without stopping at a gas station?

The answer to the previous problem comes easy if we know where we got the data from: if the information was from the US, we know we can fit inside 56.78 liters of gas, while if it comes from the UK, it is 68.19 - a difference of about 11 liters, with which a car might go for another 100 miles (or 161 km if the driver is not American or British).

V. Christophides et al. (Eds.): SWDB-ODBIS 2007, LNCS 5005, pp. 77–97, 2008.
© Springer-Verlag Berlin Heidelberg 2008

Many more such examples exist (see [NAS] for a particularly costly one) and the reason for which they persist is mainly because it is hard to change the schema of relational databases that do not include the units of their measurements simply because when they were designed, they were designed for a single context, where everybody would know what the units are. With globalization, off-shoring, out-sourcing and all the other traits of the modern economical environment, those assumptions become an obstacle to conducting efficient business processes.

Even in the context of a purely-semantic web application, such as the Potluck [pot] project developed at the Computer Science and Artificial Intelligence Laboratory at MIT (CSAIL), contextual information is not explicitly approached. The user is allowed to mash-up together information from different sites, but it is not taken into account the fact that those different data sources may have different assumptions about an entire array of concepts. This paper shows how the COIN strategy can be implemented in this new environment and how it can contribute to it.

1.1 Semantic Web ⇄ COntext INterchange

Our current work acknowledges the successes that the Semantic Web community has achieved, particularly in the standardization of expressive new languages, and builds on top of that, providing methods to address the problem of *context mediation* or *context interchange*. The two areas, Semantic Web research and Context Mediation research, are complementary to each other. Each one provides the means for, and, at the same time, enhances the other. In particular, context mediation research helps resolve semantic heterogeneity in OWL/RDF/XML data, while semantic web research provides the standards for ontology representation and reasoning. Figure 1 depicts this mutually beneficial environment.

1.2 Approach Overview

Semantic web tools rely heavily on mathematical logic to perform inferences. The result of this, in combination with our desire to maintain a 100% pure logic approach, is that *facts* cannot be deleted or modified. For instance, even if we define a relation *hasName* to be of functional type (i.e. have a unique object for each subject), it is still legal to have two entries with different objects, such as (location1,hasName,'London') and (location1,hasName, 'Londres'), the conclusion of which will be that the names "London" and "Londres" denote the same location. This has both advantages and disadvantages which we will not discuss here, but refer the reader to a wealth of literature on mathematical logic. Instead, we focus on how we model context in this framework.

Continuing the automobile-related example from the previous section, let us imagine a scenario where we are a British individual looking to purchase a car, and one of our main concerns is the environment, so we want a car that has a low gas consumption. Of course, we would prefer a more sporty car, if possible.

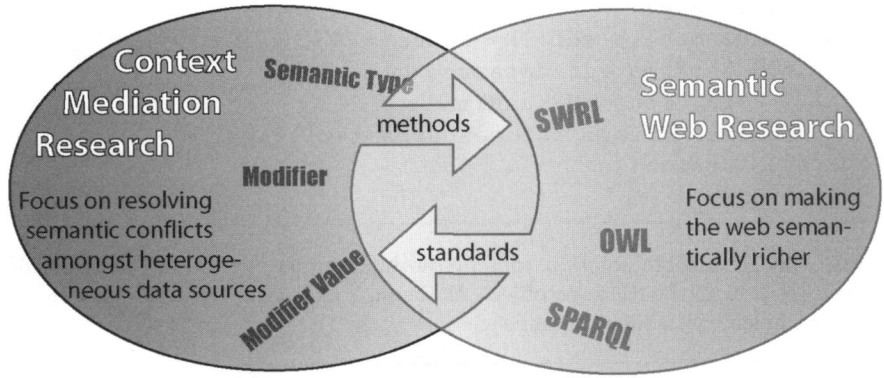

Fig. 1. Interaction between Semantic Web research and Context Mediation Research

We use the wealth of information on the Internet to do this, either via a mash-up tool like Potluck [pot], or just by browsing different car manufacturers' websites.

Imagine we look at a Ford Focus on `www.ford.com`, a mid-size car, and see that it does 24 mpg (miles per gallon). Since we are also interested in sports cars, we look at the Lotus, on `www.grouplotus.com`, and see that an introductory model does 25 mpg. We might then be convinced to spend the extra money to save the environment while enjoying the thrills of a true sports car. Unfortunately for us, it's not quite like that. We have the same misunderstanding that we mentioned before: the gallon. While the Focus considers an American gallon, the Lotus uses the British gallon. So if we transform the 24mpg of the Focus into British gallons, we get 28.8mpg - tempering our enthusiasm for the sports car.

This example is anecdotal but at the same time characteristic of the problems that occur when bringing together, mechanically, information residing in different databases.

We had previously analyzed a "Weather" example, where temperature units were converted automatically between Celsius and Fahrenheit [LM07]. Here, we extend that example and add a twist to it: now, the units are no longer different in their notation: *mpg* simply means different thing if we are in the US or in the UK.

Listing 1.1 shows an existing piece of information regarding the gas consumption of a car, while Listing 1.2 shows how we might represent different values of the same mileage using prefixes of relations.

Listing 1.1. Existing data regarding the gas consumption of a car

```
1   <?xml version="1.0"?>
2   <rdf:RDF
3     <US:mileage rdf:ID="mileage1">
4       <US:hasValue rdf:datatype="[...]#float">
5       24</US:hasValue>
```

```
6    </US:mileage>
7    <US:Automobile rdf:ID="Focus">
8      <US:hasName rdf:datatype="[...]#string">
9        Ford Focus</US:hasName>
10     <US:hasMileage rdf:resource="#mileage1"/>
11   </US:Automobile>
12 </rdf:RDF>
```

Listing 1.2. A possible solution to representing context, by adding an additional *hasValue* relation to the `Mileage` object. It is simple and intuitive of the fact that we are dealing indeed with the same mileage.

```
1  <?xml version="1.0"?>
2  <rdf:RDF
3    <US:mileage rdf:ID="mileage1">
4      <US:hasValue rdf:datatype="[...]#float">
5      24</US:hasValue>
6      <UK:hasValue rdf:datatype=''[...]#float''>
7      28.8</UK:hasValue>
8    </US:mileage>
9    <US:Automobile rdf:ID="Focus">
10     <US:hasName rdf:datatype="[...]#string">
11       Ford Focus</US:hasName>
12     <US:hasMileage rdf:resource="#mileage1"/>
13   </US:Automobile>
14 </rdf:RDF>
```

The way to add this new relation is by defining a SWRL [HPSB+04] rule such as the one in Listing 1.3 and then querying the results with a query in SPARQL [PS07] as in Listing 1.4.

Listing 1.3. SWRL rule that generates a value in the UK context

```
1  US:hasValue(?mileage, ?mileageValue) ∧
2  US:hasMileage(?car, ?mileage) ∧
3  swrlb:multiply(?product,?mileageValue,12) ∧
4  swrlb:divide(?newValue,?mileageValue,10)
5  →  UK:hasValue(?mileage, ?newValue)
```

Listing 1.4. SPARQL query that retrieves the value in the UK context

```
1  SELECT ?mileageValue
2  WHERE { ?1 US:hasMileage ?mileage .
3            ?1 US:hasName "Ford Focus" .
4            ?mileage UK:hasValue ?mileageValue
5          }
```

The SWRL rule in Listing 1.3 simply states that in order to get the UK mileage from the US mileage we have to multiply the original value by 1.2. Since SWRL does not handle floating point values, we do this my a multiplication in line 3 and a division in line 4.

The query in Listing 1.4 first identifies an entity ℓ (names starting with ? represent variables) which has a particular name ("Ford Focus") and a mileage. It returns the UK value of the mileage of the object ℓ.

This all seems very intuitive. As always, the problems lie in the details: How do we determine a conflict of contexts? How do we identify the correct rule to be applied? How should the data and rules be organized into files? Do we apply the rule to the entire dataset thus generating massive amounts of new data, or should we just apply it to the subset being queried?

After providing some background in Sections 2 and 3, we introduce the basic representation of context using the Ontology Web Language (OWL [MvH04]) in Section 4 and present our conflict identification and resolution method in Section 5.

2   Background and Related Work

2.1   COntext INterchange

The idea of the COntext INterchange [GBMS99] system is to re-use massive amounts of data that already exist but that are incomplete due to design assumptions that omitted constants from the dataset. When two or more such datasets are put together, or queried together, what were implied constants in each of them become variables in the aggregated dataset and consequently needs to be added back in the data. This is in most cases unfeasible due to the rigidity of the data structures or simply due to the fact that the end user has no control over the repository where the data exists.

The core of the COntext INterchange approach is a context mediator that rewrites queries coming from a user context into a context-sensitive mediated query that addresses the differences in meaning between the receiver and the sources. Conceptually, the context mediator is structured around a *domain model* that consists of *semantic types*, *attributes* and *modifiers*.

A semantic type is, as the name indicates, a conceptual entity. For instance, in the *Automobile* example of Section 3, the column `mileage` has no meaning by itself until it is associated with a semantic type *Mileage*. The coincidence in names is just because humans created both entities, but it should be clear to the reader that the column could very well have been named `milespergallon` and the semantic type *ST2842*. The important difference between semantic types and the columns with which they are associated is that the semantic types come enriched with semantics and attributes. In this simple example, *Mileage* has only one attribute, *value*.

In turns, an attribute may come endowed with a modifier. Again, using the *Automobile* example, we can imagine that the *value* attribute of the *Mileage* semantic type has a modifier *unit*. This is called a *modifier* because it changes the meaning of the attribute to which it refers - it modifies it. The *modifier value* could be *Kilometers per liter (kml)* or *Miles per gallon (mpg)*. In our example, we will only consider *mpg*, but keep in mind that it will mean different things in different contexts - something that we will need take into consideration and to model.

COIN uses this architecture to automatically determine differences in contexts and resolve queries in a way that is easy to interpret correctly by the user, even if the data is expressed in a different context. Existing application include financial reporting and analysis, airfare and car-rental aggregators, etc. [Fir03].

2.2   A Glance at the Semantic Web

Though the COIN methodology precedes the Semantic Web, and despite the many similarities in objectives and motivation, the two have developed mostly independently. The wide spectrum of tools that have been proposed by different research groups to achieve the targets of the original paper by Berners-Lee et al. [BLHL01] make a quick but complete summary virtually impossible. In this section we just look at the few tools that we identify to be "best", both in terms of the appropriateness for our own purposes, and also in terms of their acceptance and popularity within the Semantic Web community.

Clearly, one of the pillars of current Semantic Web research is the Web Ontology Language (OWL) [MvH04]. To query the data stored in OWL format, one could map it back to a relational database and query it with SQL or use a "native" query language such as SPARQL [PS07]. For most purposes, SPARQL can be translated back to SQL, but the advantage of it lies in being able to query directly the RDF graph that underlies any ontology. It has the status of *Working Draft* of the W3C since October 2006. The necessity of defining a new query language for tuples, such as SPARQL may be questionable at first glance, since SQL is also working with tuples, though represented in a different way, and XQuery, also developed within the W3C, addresses the problem of querying XML, of which OWL is but a flavor. In [Mel06], the author argues that though it is true that most data could be represented conceptually in RDF and expressed in either relational databases or basic XML and thus queried by either SQL or XQuery, SPARQL provides a much easier way of querying the RDF graph, making the entire development process, including debugging, more fluent.

After representation and query languages, the Semantic Web framework requires a *rule language* to make inferences on the existing data, thus enabling the creation of the smart agents described in the original Berners-Lee paper. Though SWRL [HPSB^{+}04] has gained most attention in the past few years, the language has not yet been standardized by the W3C and many different implementations exist, that rarely support the full specification, mainly because in that case the reasoning becomes undecidable. One of the most popular implementation is SWRLTab [swr] - an extension to the Protégé framework [pro], that uses mainly the Jess [jes] inference engine (though it could use other engines too). Other implementations are: R2ML [r2m], Bossam [bos], Hoolet [hoo], Pellet [pel], KAON2 [kao] and RacerPro [rac].

3 From Tables to Information

The first step towards making the data understandable by different agents[1] performing their activities in different contexts is to understand the fact that we are only dealing with representations of concepts and facts. As we exemplified before, $15 = 56.78 = 12.49$ if one is gallons (US), one is liters and one is gallons (UK). Consequently, it makes more sense to have an abstract concept representing this volume and attach to it the knowledge that it may be expressed in different ways.

A tempting way of moving information out of the restrictive relational database is to encode it using XML. Using a naïve method implemented in most database systems, this would result in, literally, a data dump. For instance, a simple table containing cars and mileage values (Table 1) can be expressed as in Listing 1.5.

Using XML does not solve our problem. As discussed in [Mad01], XML is not a silver bullet - it is just another way to express the data. It only provides a more flexible way, allowing us to add more meaning to it. A simple "data dump" from the relational database is not enough for two reasons: First, as we see in Listing 1.5, the file mixes together the structure of the data with the data itself. Conceptually, these are different and should be represented as such. Second, the data itself is stored as if to preserve the physical appearance of the table (i.e. a sequence of rows, each with a few columns) rather than to preserve its underlying meaning. It is thus clear that a different approach is needed.

Table 1. Sample relational table

Automobile	Mileage
Ford Focus	24
Lotus Elise S	25

Listing 1.5. XML representation of relational database

```
1   <?xml version="1.0"?>
2   <mysqldump xmlns:xsi="[...]XMLSchema-instance">
3   <database name="test">
4    <table_structure name="US cars">
5     <field Field="automobile" Type="varchar(20)" />
6     <field Field="mileage" Type="float(11)"  />
7    </table_structure>
8     <table_data name="cars">
9      <row>
10      <field name="automobile">Ford Focus</field>
11      <field name="mileage">24</field>
```

---

[1] we prefer the term 'agents' to show that they can be either human end-users or other computer systems.

```
12        </row>
13    [...]
14      </table_data>
15      </database>
16    </mysqldump>
```

There exist attempts to extract ontological information from relational tables [Ast04, LM04]. What we want here is not nearly as ambitious as in these works. For our purpose, we don't necessarily need to infer a full scale ontology from the data, but simply to express things that are the same as being the same and things that are different as being different. It sounds simple for a human being, but computers have serious difficulties in performing even this simple task.

The first thing we want to do is separate the structure of the representation from the data itself. Listing 1.6 shows how we can define an ontological structure to organize the data in the table. We use the term "ontological" simply because we use the ontology specification language, OWL, but one should not imagine a complex theory behind it: in this listing we simply state that we deal with two concepts (*Automobile* and *Mileage*) who are connected by a relationship *hasMileage*. The difference between this approach and the simple XML dump is that here *automobile* and *mileage* are regarded as concepts, rather than fields in a table. It is a subtle, but essential difference. Here, a particular instance of the *Automobile* class has a name, but is separate from its name. This distinction will allow us later to specify that *Ford Focus* is the same car as *Focus* and that 28.8 is the same mileage as 24 (one using British gallons and one using American gallons). This way, in the data file shown in Listing 1.7 we can define the abstract mileage *mileage1* and give it a value and then define the abstract car *Focus* and give it a name and associate it with the abstract mileage value. In these listings, an `ObjectProperty` relates two instances of two classes, while a `DatatypeProperty` relates the instance of a class to a pre-defined type (integer, string, etc.).

Listing 1.6. `legacyUS.owl`:Ontology structure for the information in the relational database

```
1     <?xml version="1.0"?>
2     <rdf:RDF[...]
3       xml:base="legacyUS.owl">
4       <owl:Ontology rdf:about=""/>
5       <owl:Class rdf:ID="Car"/>
6       <owl:Class rdf:ID="Mileage"/>
7       <owl:ObjectProperty rdf:ID="hasMileage">
8         <rdfs:domain rdf:resource="#Car"/>
9         <rdfs:range rdf:resource="#Mileage"/>
10      </owl:ObjectProperty>
11      <owl:DatatypeProperty rdf:ID="hasValue">
```

```
12    <rdfs:range rdf:resource="[...]#float"/>
13    <rdfs:domain rdf:resource="#Mileage"/>
14   </owl:DatatypeProperty>
15   <owl:FunctionalProperty rdf:ID="hasName">
16    <rdfs:range rdf:resource="[...]#string"/>
17    <rdfs:domain rdf:resource="#Car"/>
18    <rdf:type rdf:resource="#DatatypeProperty"/>
19   </owl:FunctionalProperty>
20  </rdf:RDF>
```

Listing 1.7. `legacyUSdata.owl`:Data represented using the ontological structure

```
1  <?xml version="1.0"?>
2  <rdf:RDF [...]
3    xmlns:US="legacyUS.owl#"
4    xmlns:contexts="contexts.owl#"
5    xml:base="legacyUSdata.owl">
6  <owl:Ontology rdf:about="">
7  <owl:imports>
8   <rdf:Description rdf:about="legacyUS.owl">
9   </rdf:Description>
10 </owl:imports>
11 </owl:Ontology>
12   <US:Mileage rdf:ID="mileage1">
13     <US:hasValue [...]">24</US:hasValue>
14   </US:Mileage>
15   <US:Car rdf:ID="Focus">
16   <US:hasName [...]>Ford Focus<US:hasName>
17     <US:hasMileage rdf:resource="#mileage1"/>
18   </US:Car>[...]
19 </rdf:RDF>
```

Listings 1.6 and 1.7 show the kind of input our system considers as *source*. We call the files *legacy* because they are obtained directly from existing data, without any context information. With respect to the amount of reasoning necessary at this point, our requirements are quite low since the machine needs not understand the concepts, but merely identify them as concepts rather than rows or columns in a table. The translation from the relational-model representation to our ontological representation is easily done automatically using one of the several available transformation languages such as XSLT [Cla99], FleXML [Ros01] or HaXml [WR99]. Subsequently, we can use expressions like `isSemanticType(Car, ST398)` to express the fact that the type `Car` defined in Listing 1.6 represents the conceptual type *ST*398. We refer to such expression as *elevation axioms* because they elevate the class `Car` from its meaning as a collection of entities in a legacy database, to a conceptual level. Using such elevation axioms instead of attaching properties to the original class Car reduces the amount of work that the user needs to do by increasing the reusability of the code.

4   Separating Context from Data Representation

In order to be able to do the things outlined in Section 1.2 we first need to establish a way to represent context. The flexibility of the RDF and OWL languages allow for such a variety of architectures to be defined, that one of the problems we faced was focusing on one in particular, one that provides, in our opinion, the best solution for future extensions.

Initially, we had reduced the possibilities to three models (Figure 2).

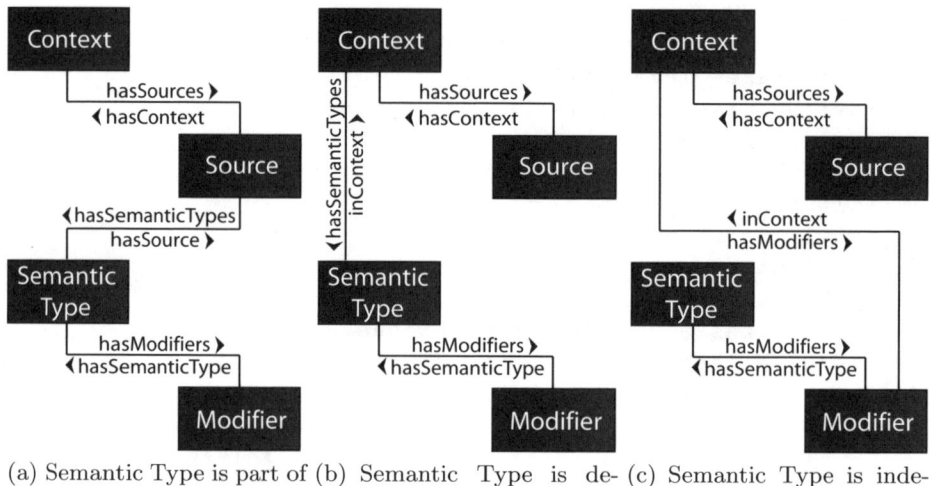

(a) Semantic Type is part of the Source

(b) Semantic Type is defined in the Context

(c) Semantic Type is independent of the Context

Fig. 2. Models for context expression

4.1   Model 1: Semantic Type as Part of the Source

This model was our initial approach and it states that a data source should have a context and that it should contain a set of *semantic types*. It is the most basic approach because it attempts to link everything directly to the legacy data source.

This model was not eventually acceptable because a source should not actually contain a *semantic* object. It contains objects that we subsequently identify as being automobiles, mileages, temperatures, locations, or anything else. By itself, it contains only some non-identifiable classes and objects that, in the COIN methodology, have to be related to semantic types via the elevation axioms.

4.2   Model 2: Semantic Type as Part of the Context

From the first model, we have learned that the semantic types need to be defined separately from the data itself. Consequently, we considered having them defined as part of the context. This method provides sufficient flexibility to allow each user that defines his or her own context to have complete freedom as to what it considers to be significant types and how these should be represented.

The disadvantage of the method also lies in the flexibility we just mentioned: additional mediation is needed and even if two users define two contexts with semantic types having exactly the same representation, they still appear as duplicates when everything is put together to allow query answering. (see Figure 3: the instance browser at the middle of the image shows duplicate semantic types corresponding to each of the two contexts defined)

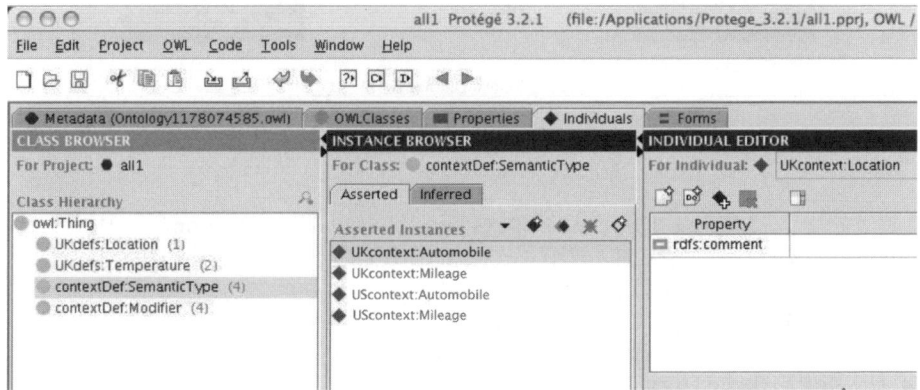

Fig. 3. Duplicate definitions of Semantic types using the second model

4.3   Model 3: Semantic Types Defined Independently of Everything Else

Finally, the chosen model considers the semantic types to be independent of both the context and the source. In fact, we should imagine these semantic types as defined in an external ontology. This method provides the most independence between the different concepts. Figure 4 presents a more detailed view than the one in Figure 2, showing the files used for each component and referencing Listings presented in this paper. We will be using three files to express context, in addition to the two files that represent the data. Two of the three are also represented in Figure 4, while the third - the `contextDefs.owl` file in Listing 1.8 provides the basic definitions needed for context representation. As such, it can be though of as the entire Figure 4 (without the sources). The `contextDefs.owl` files defines the context as a set of modifiers attached to a semantic type (some items present in the file will be explained in the next sections).

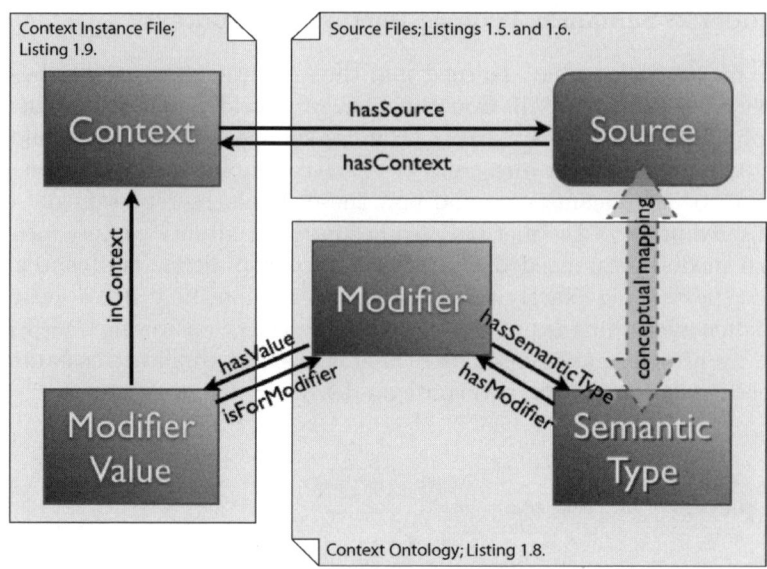

Fig. 4. Our model, along with the files that contain each component

Listing 1.8. `contextDefs.owl`: the context definition

```
1   <?xml version="1.0"?>
2   <rdf:RDF
3     xml:base="contextDefs.owl">
4     <owl:Ontology rdf:about=""/>
5     <owl:Class rdf:ID="Context"/>
6     <owl:Class rdf:ID="Query"/>
7     <owl:Class rdf:ID="TriggeredRules"/>
8     <owl:Class rdf:ID="SemanticType"/>
9     <owl:Class rdf:ID="Modifier"/>
10    <owl:Class rdf:ID="ModifierValue"/>
11    <owl:ObjectProperty rdf:ID="hasModifiers">
12      <rdfs:range rdf:resource="#Modifier"/>
13      <rdfs:domain rdf:resource="#SemanticType"/>
14    </owl:ObjectProperty>
15    <owl:ObjectProperty rdf:ID="isSemanticType">
16      <rdfs:range rdf:resource="#SemanticType"/>
17    </owl:ObjectProperty>
18    <owl:DatatypeProperty rdf:ID="hasValue">
19      <rdfs:range rdf:resource="[...]#string"/>
20      <rdfs:domain rdf:resource="#ModifierValue"/>
21    </owl:DatatypeProperty>
22    <owl:ObjectProperty rdf:ID="hasContext"/>
23     <owl:ObjectProperty rdf:ID="inContext">
```

```
24    <rdfs:range rdf:resource="#Context"/>
25    <rdfs:domain rdf:resource="#ModifierValue"/>
26  </owl:ObjectProperty>
27   <owl:ObjectProperty rdf:ID="isForModifier">
28    <rdfs:range rdf:resource="#Modifier"/>
29  </owl:ObjectProperty>
30    <owl:DatatypeProperty rdf:ID="ruleName">
31    <rdfs:domain rdf:resource="#TriggeredRules"/>
32    <rdfs:range rdf:resource="[...]#string"/>
33  </owl:DatatypeProperty>[...]
34  </rdf:RDF>
```

The following listing (Listing 1.9) contains the file that represents the system's *understanding of the world*: a list of concepts, what properties they have (modifiers) and how they relate to each other. In our example, it states that automobiles have a *mileage* attribute that is measured by *mileage unit*.

Listing 1.9. `contextOntology.owl`: Semantic types definitions (including the modifiers they accept)

```
1  <?xml version="1.0"?>
2  <rdf:RDF
3   xmlns="contextOntology.owl#"
4   xmlns:contextDef="contextDefs.owl#"
5   xml:base="contextOntology.owl">
6  <owl:Ontology rdf:about="">
7   <owl:imports rdf:resource="contextDefs.owl"/>
8   </owl:Ontology>
9   <contextDef:SemanticType rdf:ID="Mileage">
10    <contextDef:hasModifiers>
11     <contextDef:Modifier rdf:ID="MileageUnit"/>
12    </contextDef:hasModifiers>
13   </contextDef:SemanticType>
14   <contextDef:SemanticType rdf:ID="Automobile"/>
15  </rdf:RDF>
```

Finally, Listing 1.10 shows the instance of a context: all, or just a subset of modifiers, are assigned values in this file. In this listing, Lines 10-13 give a label to the context, which will be used in differentiating modifier values with similar representations in different contexts (like *mpg* in our case). Then, lines 14-20 define the *mpg* modifier value, identifying its context and the modifier it applies to.

Listing 1.10. `UScontext.owl`: Context instance file

```
1  <?xml version="1.0"?>
2  <rdf:RDF
```

```
3    xmlns="UScontext.owl#"
4    xmlns:contextOntology="contextOntology.owl#"
5    xmlns:contextDef="contextDefs.owl#"
6    xml:base="UScontext.owl">
7    <owl:Ontology rdf:about="">
8     <owl:imports rdf:resource="contextOntology.owl"/>
9    </owl:Ontology>
10   <contextDefs:Context rdf:id="USContext">
11     <rdfs:label rdf:datatype="[...]#string">
12       USContext</rdfs:label}
13   </contextDefs:Context>
14   <contextDefs:ModifierValue rdf:ID="mpg">
15    <contextDefs:inContext rdf:resource="#USContext"/>
16    <contextDefs:isForModifier rdf:resource=
17     "contextOntology.owl#MileageUnit"/>
18    <contextDefs:hasValue rdf:datatype="[...]#string">
19       mpg</contextDefs:hasValue>
20    </contextDefs:ModifierValue>
21  </rdf:RDF>
```

4.4   Automatic Identification of Context

In Listing 1.10 we identified explicitly the context of a modifier value by means of the inContext property in line 15. This may seem redundant considering that the value is defined in a file specific to this context. In fact, this property can be automatically asserted using the from named construct in the SPARQL query language. Listing 1.11 shows how this can be done, with results in Table 2.

Listing 1.11. Using the from named construct we can identify to which context each modifier value belongs to

```
1  prefix contextDefs:<contextDefs.owl#>
2  select ?src ?modifier ?value
3  from named <UKcontext.owl>
4  from named <UScontext.owl>
5  where {
6   graph ?src{
7    ?modifier contextDefs:hasValue ?value
8   }
9  }
```

The results in column *src* of Table 2 may be used to replace the labels in Listing 1.10. In the remaining of the presentation we will assume that the labels have already been set, either manually, or using the method we just indicated.

Table 2. Result of query in listing 1.11

src	modifer	value
contextUS.owl	contextOntology:MileageUnit	mpg
contextUK.owl	contextOntology:MileageUnit	mpg

4.5 The Mediator File

To make the system work, one file needs to import all these bits and pieces together and build the construct of the COntext INtegration strategy. We call this the *mediator* file. Listing 1.12 shows an extract of its contents, in particular the `import` statements and the way we define the context of a source. In this example, the source is just the Mileage entity `mileage1` defined in the `legacyUSdata.owl` file (Listing 1.7). However, it can be anything else: an entire file, a class or just an instance as in this case.

Listing 1.12. The *mediator* file puts together all the different pieces of the architecture and defines particular contexts of the sources

```
1  <rdf:RDF>
2    xmlns:UKdefs="legacyUK.owl#"
3    xmlns:USdefs="legacyUS.owl#"
4    xmlns:USdata="legacyUSdata.owl#"
5    xmlns:contextDefs="contextDefs#"
6    [...]
7    <owl:Ontology rdf:about="">
8      <owl:imports rdf:resource="UScontext.owl"/>
9      <owl:imports rdf:resource="UKcontext.owl"/>
10     <owl:imports rdf:resource="legacyUSdata.owl"/>
11     <owl:imports rdf:resource="legacyUK.owl"/>
12   </owl:Ontology>
13   <rdf:Descripton rdf:about="legacyUSdata.owl#mileage1">
14     <j.0:isSemanticType rdf:resource=
15     "contextOntology.owl#Mileage"/>
16     <j.0:hascontext rdf:resource="UScontext.owl#"/>
17   </rdf:Description>
18 </rdf:RDF>
```

In the listing above, lines 2-4 give names to particular ontologies, names which we will use in defining the rules and the queries below.

Now we can, for instance, identify the context of a source using a query similar to the following:

Listing 1.13. Query to identify the context of a source

```
1  prefix contextDefs:<contextDefs.owl#>
2  select ?data ?context
3    where {?data contextDefs:hasContext ?context}
```

In our example, the results of this query is shown in Table 3.

Table 3. Results of the context query in Listing 1.13

data	context
USdata:mileage1	USContext:USContext

5   Context Conflict Identification and Resolution

In the previous sections we have explained how a user might query the data to find out what is the appropriate context it refers to. In this section we will show how we do this automatically for the purpose of context conflict determination and how this determination will trigger the necessary conversion rules. We will continue to use the example of cars and mileages presented throughout this work.

The approach we follow in this work is a two-step approach: first, we need to determine the need for a conversion (i.e. determine the existence of two different contexts) and then apply the corresponding rule.

These two phases are implemented in two sets of rules: first a *trigger rule* analyses the data and the query to identify potential conflicts. If one such conflict is identified, a flag is raised, to announce the necessity of the application of a conversion rule. We will describe the implementation of this flag shortly. Upon assertion of the trigger flag, the corresponding rule will automatically be triggered and context mediation will take place by addition of new data to the dataset.

5.1   The Trigger Rule

The idea of the trigger rule is to look for conflicts and add a flag, in the form of a small text representing the needed conversion, which is added to a collection of triggers called `TriggeredRules1` in our example.

Listing 1.14 shows the exact rule used to determine conflicts between any two attributes of the same type.

Listing 1.14. Rule for the determination of context conflict

```
 1  USdefs:hasValue(?attribute, ?attributeValue) ∧
 2  contextDefs:hascontext(?attribute, ?dataContext) ∧
 3  contextDefs:hascontext(Query_1, ?queryContext) ∧
 4  differentFrom(?dataContext, ?queryContext) ∧
 5  contextDefs:isSemanticType(?temp, ?semType) ∧
 6  contextDefs:hasModifiers(?semType, ?modifier) ∧
 7  contextDefs:isForModifier(?modVal, ?modifier) ∧
 8  contextDefs:hasValue(?modVal, ?dataModVal) ∧
 9  contextDefs:inContext(?modVal, ?datacontext) ∧
10  contextDefs:isForModifier(?modVal1, ?modifier) ∧
11  contextDefs:hasValue(?modVal1, ?queryModVal) ∧
12  contextDefs:inContext(?modVal1, ?querycontext) ∧
13  rdfs:label(?queryContext, ?c1)   ∧
```

```
14   rdfs:label(?dataContext, ?c2)  ∧
15   swrlb:stringConcat(?tn0,":",?queryModifierValue)  ∧
16   swrlb:stringConcat(?tn1,?c1,?tn0)  ∧
17   swrlb:stringConcat(?tn2,"-to-",?tn1)  ∧
18   swrlb:stringConcat(?tn3,?dataModifierValue,?tn2)  ∧
19   swrlb:stringConcat(?tn4,":",?tn3)  ∧
20   swrlb:stringConcat(?triggername,?c2,?tn4)
21     → contextDefs:ruleName(TriggeredRules1, ?triggername)
```

A detailed explanation follows:

Lines 1-4 identify the difference between the contexts of the data and the query.
Lines 5-6 identify the semantic type and modifier
Lines 7-9 identify the value of the modifier in the context of the dataset
Lines 10-12 do the same for the value of the modifier in the query context.
Lines 13-14 identify the labels of each context, used later in generating the trigger name
Lines 15-20 generate the name of the trigger
Line 21 asserts the trigger

In this implementation, the attribute itself is linked by a `hasContext` relation to a particular context. In other situations, such a relation may only be defined for the entire dataset, rather than for individual attributes. This is not a problem, as a SWRL rule can extend the `hasContext` rule from a class to its components.

In our running example, this rule would generate a flag of the form *USContext: mpg-to-UKContext:mpg.*

5.2   The Conversion Rule

The actual conversion rule that transforms the miles per gallon measure unit from Winchester gallons to Imperial gallons is shown in Listing 1.15. Line 1 checks the existence of the triggered flag and, if this condition is satisfied, it performs the necessary mathematical conversion functions (lines 2-4) and asserts the new value in line 5.

The result of applying both rules on the knowledge base is shown in Figure 5. We can see that two new facts have been asserted: first, the trigger rule has discovered the context conflict and, second, upon assertion of the conflict, the conversion rule has been triggered to compute the new value.

Listing 1.15. Conversion rule

```
1   contextDefs:ruleName(TriggeredRules1,
2       "USContext:mpg-to-UKContext:mpg")  ∧
3   USdefs:hasValue(?mileage, ?mileageValue)  ∧
4   swrlb:multiply(?mileage1, ?mileageValue, 12)  ∧
5   swrlb:divide(?newValue, ?mileage1, 10)
6     →   UKdefs:hasValue(?mileage, ?newValue)
```

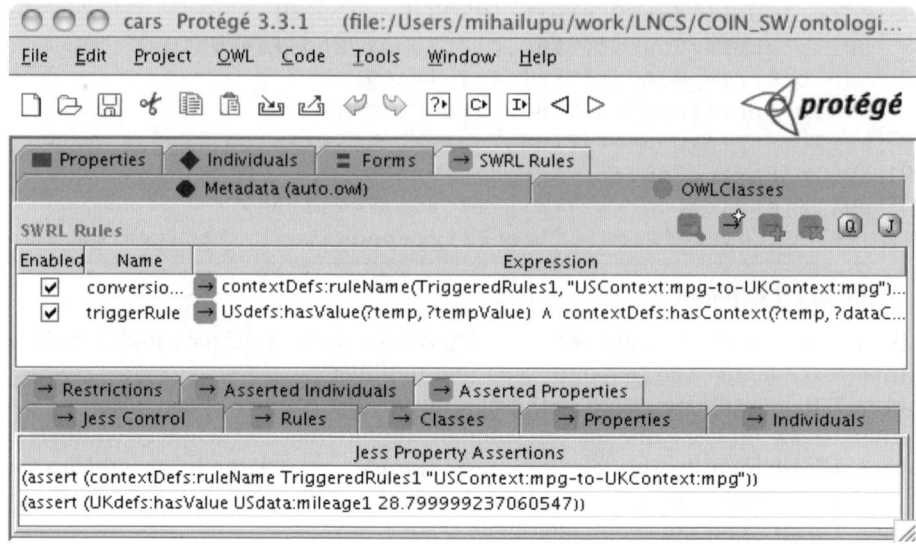

Fig. 5. With two rules in the knowledge base, the conversion between mileage units has been performed automatically. In the lower half of the image we can see that the new mileage value (28.8) has been correctly asserted.

5.3 Query Alteration

The simple way in which we have defined the conversion rule (Listing 1.15) allows us to re-write the query in an equally simple manner. Listings 1.16 and 1.17 show the original and, respectively, the new query for obtaining the value of the mileage for the Focus. As it can be observed, the only difference is the prefix of the hasValue relationship.

Listing 1.16. Original query

```
1  SELECT ?mileageValue
2  WHERE { ?loc USdefs:hasMileage ?mileage.
3          ?loc USdefs:hasName "Ford Focus".
4          ?mileage USdefs:hasValue ?mileageValue}
```

Listing 1.17. New query for the UK context

```
1  SELECT ?mileageValue
2  WHERE { ?loc USdefs:hasMileage ?mileage.
3          ?loc USdefs:hasName "Ford Focus".
4          ?mileage UKdefs:hasValue ?mileageValue}
```

Despite its ease of use and implementation, the current method is not perfect. The relation UKdefs:hasValue is not directly linked to the UK context.

Formally, it has no link to any specific context. Though this approach can be implemented programatically, our future work, described in the next section, aims towards an ever closer integration with the semantic web tools.

6   Future Work

Our work so far has shown how we can approach the problem of context interchange using the COIN strategy via the tools of the Semantic web. To fully achieve all the features that are currently available in COIN there are still steps ahead, some of which we describe in this section.

The solution presented in the previous section relies on external programing languages to transform a query such that it returns the result in a different context. A better solution would be to have a new tertiary relation, similar to the one that defined the value of a modifier in a particular context. This new relation, which we call `hascontextValue` links together an attribute, a value and a context. As we have seen, SWRL can only express binary relations directly, so the only way to implement this relation is to define it as an owl:class with three binary relations. Now, the conversion rule needs to infer the new tertiary relation that links the attribute to the new value in the new context. Such a rule can be created following the Semantic Web best practices [NRHW06] as in Listing 1.18.

Listing 1.18. Tertiary relation implemented as an OWL class

```
1  <owl:Class rdf:ID="hascontextValueRelation"/>
2    <owl:ObjectProperty
3       rdf:ID="hascontextValueRelation_context">
4      <rdfs:range rdf:resource="#Context"/>
5    </owl:ObjectProperty>
6    <owl:ObjectProperty
7       rdf:ID="hascontextValueRelation_attribute"/>
8    <owl:ObjectProperty
9       rdf:ID="hascontextValueRelation_value"/>
```

The difficulty in inferring this relation in the conversion rule is that a new instance has to be generated: a new individual of the `hascontextValue` type that would link the three components (attribute, value and context). Unfortunately, the current standard SWRL specification does not provide means to instantiate classes, thus making this solution temporarily unfeasible. This leavs only the option of an "impure" approach using external programming tools.

7   Conclusion

In this work we describe how the COntext INterchange strategy can be implemented using the Semantic Web tools, in particular using OWL, SWRL and

SPARQL. We acknowledge the existence of massive amounts of data in relational databases that lack all the necessary data required for users other than the original designers of the database and describe how the information present in these databases can be "elevated" to a knowledge base. Subsequently, we show how to structure information pertaining to the context of the data - how to model the definitions of *semantic type, modifier* and *modifier value*. Using these models we show how the necessary conversions of the data values can be made by using a two-step process involving pairs of *trigger* and *conversion* rules.

References

[Ast04]   Astrova, I.: Reverse Engineering of Relational Databases to Ontologies. In: Bussler, C.J., Davies, J., Fensel, D., Studer, R. (eds.) ESWS 2004. LNCS, vol. 3053, pp. 327–341. Springer, Heidelberg (2004)

[BLHL01]   Berners-Lee, T., Hendler, J., Lassila, O.: The semantic web. Scientific American (2001)

[bos]   Bossam, http://bossam.wordpress.com/

[Cla99]   Clark, J.: Xsl transformations (1999), http://www.w3.org/TR/xslt

[Fir03]   Firat, A.: Information Integration using Contextual Knowledge and Ontology Merging. PhD thesis, Sloan Business School, MIT (2003)

[GBMS99]   Goh, C.H., Bressan, S., Madnick, S., Siegel, M.: Context interchange: New features and formalisms for the intelligent integration of information. ACM TIS 17(3) (1999)

[hoo]   Hoolet, http://owl.man.ac.uk/hoolet/

[HPSB+04]   Horrocks, I., Patel-Schneider, P., Boley, H., Tabet, S., Grosof, B., Dean, M.: SWRL: A semantic web rule language combining OWL and RuleML (2004), http://www.w3.org/Submission/SWRL/

[jes]   The Jess inference engine, http://herzberg.ca.sandia.gov/jess/

[kao]   Kaon2, http://kaon2.semanticweb.org/

[LM04]   Lammari, N., Metais, E.: Building and maintaining ontologies: a set of algorithms. In: NLDB, vol. 48(2) (2004)

[LM07]   Lupu, M., Madnick, S.: Using semantic web tools for context interchange. In: Proc. of SWDB-ODBIS Workshop (2007)

[Mad01]   Madnick, S.: The misguided silver bullet: What XML will and will not do to help information integration. In: Procs. of the iiWAS (2001)

[Mel06]   Melton, J.: SQL, XQuery, and SPARQL:what's wrong with this picture? In: Proc. of XTech Conference (2006)

[MvH04]   McGuinness, D., van Harmelen, F.: Owl web ontology language overview (2004), http://www.w3.org/TR/owl-features/

[NAS]   NASA. Mars climate orbiter failure causes, http://mars.jpl.nasa.gov/msp98/news/mco990930.html

[NRHW06]   Noy, N., Rector, A., Hayes, P., Welty, C.: Defining n-ary relations on the semantic web (2006), http://www.w3.org/TR/swbp-n-aryRelations/

[pel]   Pellet, http://www.mindswap.org/2003/pellet/

[pot]   Potluck mash-up tool, http://dfhuynh.csail.mit.edu:6666/potluck/

[pro]   Protégé, http://protege.stanford.edu/

[PS07]   Prud'hommeaux, E., Seaborne, A.: Sparql query language for rdf (2007), http://www.w3.org/TR/rdf-sparql-query/

[r2m]   R2ml,
            http://oxygen.informatik.tu-cottbus.de/rewerse-i1/?q=node/6
[rac]   RacerPro,
            http://www.racer-systems.com/products/racerpro/index.phtml
[Ros01]   Rose, K.: FleXML - XML Processor Generator (2001),
            http://flexml.sourceforge.net
[swr]   SWRLTab, http://protege.cim3.net/cgi-bin/wiki.pl?SWRLTab
[WR99]   Wallace, M., Runciman, C.: Haskell and XML: Generic Combinators or
            Type-Based Translation? In: Proc. of the International Conference on
            Functional Programming (1999),
            http://www.cs.york.ac.uk/fp/HaXml/

On the Synthetic Generation of Semantic Web Schemas

Yannis Theoharis[1,2], George Georgakopoulos[2], and Vassilis Christophides[1,2]

Institute of Computer Science, FORTH, Vassilika Vouton, P.O.Box 1385, GR 71110, Heraklion, Greece
Department of Computer Science, University of Crete, P.O.Box 2208, GR 71409, Heraklion, Greece
{theohari,christop}@ics.forth.gr,
ggeo@csd.uoc.gr

Abstract. In order to cope with the expected size of the Semantic Web (SW) in the coming years, we need to benchmark existing SW tools (e.g., query language interpreters) in a credible manner. In this paper we present the first RDFS schema generator, termed PoweRGen, which takes into account the morphological features that schemas frequently exhibit in reality. In particular, we are interested in generating synthetically the two core components of an RDFS schema, namely the *property* (relationships between classes or attributes) and the *subsumption* (subsumption relationships among classes) graph. The total-degree distribution of the former, as well as the out-degree distribution of the Transitive Closure (TC) of the latter, usually follow a `power-law`. PoweRGen produces synthetic property and subsumption graphs whose distributions respect the `power-law` exponents given as input with a confidence ranging between $90 - 98\%$.

1 Introduction

Semantic Web (SW) [4] gains increasing popularity nowadays. A significant amount of RDF/S schema and instance descriptions are already available on the WWW. According to [3], 16.69% of XML documents rely on RDF/S [6], while their volume reaches the 31.71% of the available XML data. As the size of the SW is expected to be further increased in the coming years, the scalability of SW storage, query or update tools becomes crucial [24]. Following the tradition of data management community, new benchmarks and synthetic data generators need to be developed for the SW. This need is also underlined by the recent RDF/S benchmarking initiatives undertaken by the W3C consortium [28].

Unlike XML data generators [21,2], which rely on fixed DTDs or XML schemas, we pay particular attention to the synthetic generation of SW schemas specified in RDFS [6]. This is motivated by the fact that advanced SW reasoning functionalities (e.g., concerning subsumption relationships) can be built only by exploiting the additional information encoded in these schemas.

Our focus in RDFS, instead of OWL, schemas is motivated by the fact that the majority of available SW schemas rely on the RDFS specification (according

V. Christophides et al. (Eds.): SWDB-ODBIS 2007, LNCS 5005, pp. 98–116, 2008.

to [9], 85.45% relies on RDFS, while 14.55% on OWL). Additionally, the core RDFS features are also exploited by OWL. Therefore, our work can be seen as a basis that can be extended to OWL schema generation.

RDFS schemas are essentially graphs whose arcs are of different nature, namely, a) arcs representing subsumption relationships among classes, and b) arcs representing relations between classes (e.g., *has_a*) or attributes (e.g., *title*), collectively called properties. In this context, for each RDFS schema we essentially need to generate two graphs that have the same set of nodes (i.e., classes or literal types), namely, the *subsumption*, and the *property* graph.

It is well known that a `power-law` is a function of the form $f(x) \propto x^{-b}$. The *total-degree* distribution of the property graph, as well as the *out-degree* (i.e., the class descendants) distribution of the Transitive Closure (TC) of the subsumption graph usually follow a `power-law` distribution [25]. Furthermore, classes that appear as domain of many properties are located highly in the class hierarchies, i.e., classes with high out-degree in the *property* graph are located at the first levels of the *subsumption* graph. Thus, synthetic RDFS schemas should exhibit similar features with those frequently encountered in reality.

In this paper we propose the first synthetic RDFS schema generator, termed PoweRGen, which takes as input: a) the number of schema classes and properties, b) the characteristic exponents of the aforementioned `power-law`s, c) the depth of the subsumption graph, and d) the knowledge whether the subsumption graph should be a DAG or a tree. The property graph of the generated schemas follows the `power-law` given as input (98% confidence). The same is also true for the subsumption graph in the case of trees. For DAGs, the generated subsumption graph approximates with a 90.3% confidence a `power-law`, whose characteristic exponent follows (99.4% confidence) the one given as input.

The remainder of this paper is organized as follows: Section 2 introduces the main features of the *property* and *subsumption* graph forming an RDFS schema. Section 3 presents the generation of RDFS schemas. Section 4 presents the results of an experimental evaluation, while Section 5 compares our graph generation method with related work. Finally, Section 6 identifies issues for future research.

2   Semantic Web Schema Graphs

RDFS schemas are usually represented as directed labeled graphs, whose nodes are classes or literal types and arcs are properties. These graphs may have self-loops (representing recursive properties) and multiple arcs (when two classes are connected by several properties). The leftmost part of Figure 1 depicts an example of a schema. In particular, SW schemas have two different kinds of arcs: subsumption arcs (*rdfs:subclassOf*), and user defined ones. The former comprises *subsumption* relationships among classes (which are transitive in nature), while the latter comprises attributes or relationships among classes, which are called *properties*. As the interpretation of these two arc kinds is different, for each RDFS schema we need to define two graphs: a) the *property* (e.g., the second part of Figure 1) and b) the *subsumption* graph (e.g., the third part of Figure 1).

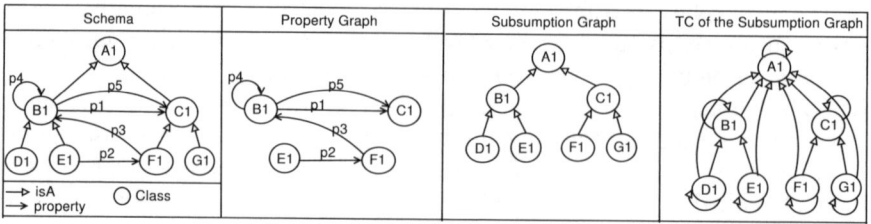

Fig. 1. A SW schema and its constituent graphs

Both graphs have the same set of nodes (i.e., the union of classes and literal types used in the schema) but they comprise different kinds of arcs.

Definition 1. *The* property graph *of a schema is a directed graph* $\mathcal{G}_p = (\{C \cup L\}, P)$, *where* C *is a set of nodes labeled with a class name,* L *is a set of nodes labeled with a literal type,* P *is a set of arcs of the form* $\langle c_1, p, c_2 \rangle$ *where* $c_1 \in C$, $c_2 \in C \cup L$, *and* p *is a property name.*

Definition 2. *The* subsumption graph *of a schema is a directed graph* $\mathcal{G}_s = (C, P_s)$, *where* C *is a set of nodes labeled with a class name and* P_s *is a binary relation over the elements of* C.

By considering the RDFS semantics [6] we can obtain the TC of the subsumption graph of a schema (e.g. the rightmost part of Figure 1).

Definition 3. *Let* $\mathcal{G}_s = (C, P_s)$ *be the subsumption graph of a schema. The* TC *of the subsumption graph, denoted by* $\mathcal{G}_s^*$, *is the pair* (C, P_s^*) *where* P_s^* *is the* TC *of* P_s.

It should be stressed that, according to the RDF/S semantics [14], class subsumption is a transitive relation and hence we should study the TC of the subsumption graph. The effect of the subsumption to the property graph is implicit.

Concerning the *property* graph, the out-(resp. in-) degree of a class is the frequency of occurrence of this class in the ranges (resp. domains) of properties, while the total-degree of a class is the frequency of occurrence of this class in either the ranges or the domains of properties. Furthermore, the out-(resp. in-) degree of a class in the TC of the *subsumption* graph corresponds to transitive subclasses (resp. superclasses) of this class.

2.1   RDFS Schema Graph Features

In this Section, we briefly recall the experimental results of [25] highlighting the main graph features exhibited by an adequate large corpus[1] of big sized schemas available on the SW. In particular, we are interested on the core RDFS features

---

[1] For more information see http://athena.ics.forth.gr:9090/RDF/VRP/SWSchemas/

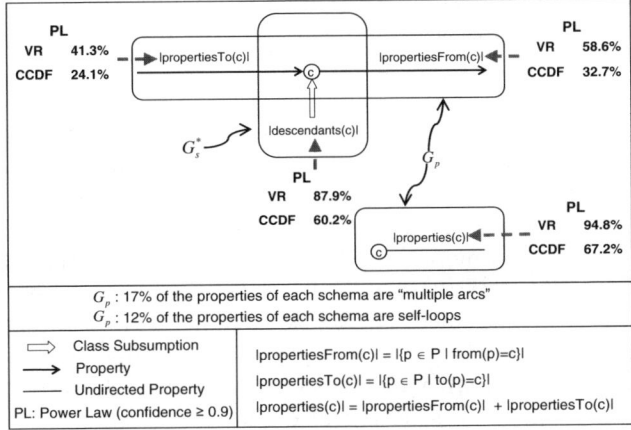

Fig. 2. Main graph features of SW schemas

which are also exploited by OWL. We report the degree distributions of the property and subsumption graph. Additionally, we report combinatorial findings concerning the two graphs (e.g., the class distribution per level of the subsumption hierarchy, as well as, the number of related properties) which allows us to sketch a more accurate picture about the morphology of real RDFS schemas.

Degree Distributions. In order to analyze the degree distributions of schema graphs, 4 (mainly) Discrete Random Variables (DRV) were considered. In particular, 3 of them corresponded to the in-/out-/total-degrees of the *property* graph, $\mathcal{G}_p$, and the fourth to the out-degrees (corresponding to class descendants) of the TC of the *subsumption* graph, $\mathcal{G}_s^*$. Possible power-laws were investigated on two functions related to each DRV:

i) complementary cumulative probability density function (CCDF), i.e., $P(X \geq x)$
ii) the relationship among the values of its range set, denoted by D, and their rank in decreasing order (VR), i.e., a function $f : [1, ..., |D|] \rightarrow D$.

Implicitly, probability density function (PDF), i.e., $P(X = x)$, was also considered, since CCDF and PDF are essentially the same and whenever PDF follows a power-law with exponent b, CCDF follows a power-law with exponent $b - 1$ [7].

The results of the analysis of [25] are summarized in Figure 2. The main conclusion drawn was that the majority of RDFS schemas approximate a power-law for the total-degree functions (94.8% for VR and 67.2% for CCDF) of the property graph, while for out- and in-degrees the corresponding percentages are significantly lower. The characteristic power-law exponents for the total-degree VR (resp. CCDF) lie in $[0.79, 2.18]$ (resp. $[0.65, 2.05]$). Concerning the subsumption graph, the out-degree VR (resp. CCDF) approximate a power-law for the 87.9% (resp. 60.2%) of RDFS schemas. The characteristic exponents in this case lie in

[0.97, 2.44] (resp. [0.54, 1.47]) for VR (resp. CCDF). Class ancestors distribution was not observed to follow a `power-law` (nor any other) distribution for real SW schemas.

Morphological Features. An interesting observation that correlates the property graph with the subsumption graph is that most properties have as domain, classes which are located highly in the class hierarchy, i.e., somewhere between the root and the middle level. This fact shows that the specification of a class in RDFS schemas is used more for classification purposes rather than for refining classes with additional properties. The same trend was observed for range classes, although the dominance of classes located higher over those located lower in the subsumption graph is not so important as for property domains.

Furthermore, the consideration of one additional DRV corresponding to number of classes per hierarchy level revealed the skewness of the distribution of classes per level. Specifically, the VR function approximates (for the 42.1% of real RDFS schemas) a `power-law` while the PDF always approximates the uniform distribution. Moreover, the corresponding exponent of the VR function approximately depends linearly on the number of nodes (classes). In addition, the level, denoted by k, at which the maximum number of nodes are located is approximately $0.75 \times depth$ for real RDFS schemas, while the number of classes located at a specific level decreases as long as its distance to k increases. Last, on average the 75% of classes of real RDFS schemas are leaves.

Finally, there exists a strong correlation between the out- and in-degrees of nodes of $\mathcal{G}_p$. In other terms, classes that appear as domain of many properties appear also as range of many (the same or other) properties. Last but not least, the percentage of self-loops (i.e., recursive properties) is quite significant, i.e., 12.6% in the average case. The average percentage of multiple arcs, that is properties that share the same domain and range, is slightly higher, i.e., 17.7%.

3   Synthetic SW Schema Generation

In this Section, we present the main algorithmic steps of the `power-law` based RDFS synthetic schema generator (PoweRGen). In particular, PoweRGen comprises the:

i) generation of the total-degree (resp. out-degree) sequence of $\mathcal{G}_p$ (resp. $\mathcal{G}_s^*$). To this end, it exploits two methods to sample a bag of values from a DRV that follows a `power-law`,

ii) generation of the property (resp. subsumption) graph by using $GenerateG_p$ (resp. $GenerateG_s$) algorithms. These algorithms are based on *Linear Programming* and exploit features that schemas frequently exhibit in reality (see Section 2.1)

iii) mapping of nodes of $\mathcal{G}_p$ to nodes of $\mathcal{G}_s^*$ in order to construct the output schema.

In the sequel, we elaborate on each one of the above three steps.

3.1   Sampling a DRV That Follows a PL

To generate the total-degree sequence of $\mathcal{G}_p$, as well as the out-degree sequence of $\mathcal{G}_s^*$. we exploit two methods (see [23]) to sample a bag of values from a DRV that follows a `power-law`. The former implements well known ideas and is independent of the nature of the PDF function. The latter applies only when the VR function is a `power-law`. The former is useful in the case that the DRV is bounded by a maximum value. This is actually the case of out-degrees of $\mathcal{G}_s^*$ in which the maximum allowed value is $|C| - 1$ (i.e., root descendants). The latter is useful in the case that the sum of the bag of sampled values is predefined. This is actually the case of the total-degrees of $\mathcal{G}_p$, where the sum of all total-degree values should be equal to twice the number of schema properties (graph arcs).

3.2   Generating the Property Graph

Bellow, we explain in detail algorithm $GenerateG_p$.

Algorithm. $GenerateG_p$

Inputs.
      b: the VR exponent of the total-degree distribution,
      N_c/N_p: the number of schema classes/properties,
      p_0: percentage of classes that neither appear as domain
         nor as range of any property.

(1) $D := VRSampling(b, \lfloor(1 - p_0) \times N_c\rfloor, 2 \times N_p)$;
      $D_{in} := \emptyset;\ D_{out} := \emptyset$;
 (2)   for $i = 1$ to N_c do
        $D_{in}[i] = D_{out}[i] = \frac{1}{2} \times D[i]$;
 (3)   Choose randomly nodes and attach on them a set S
      of $\lfloor 0.126 \times N_p \rfloor$ self-loops (modify D_{in} and D_{out});
 (4)   Choose randomly pairs of nodes and attach on them a set
      M of $\lfloor 0.177 \times N_p \rfloor$ multiple arcs (modify D_{in} and D_{out});
 (5) $E :=$ the set of arcs of the solution of the
      $LP1$ instance corresponding to D_{in} and D_{out};
 (6) $E_p = E \uplus S \uplus M$;

Using $VRSampling$ we can generate the *total-degree* sequence, denoted by D, of the $\mathcal{G}_p$ based on the VR exponent of the total-degree distribution, denoted by b (step 1). We choose to generate D instead of $Dout$ and Din, because the percentage of real RDFS schemas that approximate a `power-law` for the *total-degree* distribution is bigger than the corresponding percentages for *out-* and *in-degree* distributions (see Section 2.1). We should also mention that there exist classes in real SW schemas that neither appear as domain nor as range of any property, i.e., their total-degree is 0. The percentage of such classes, denoted by p_0, can be a parameter of the generator (a typical value is 50%). Hence, we generate D by setting $(b, N, sum) = (b, \lfloor(1 - p_0) \times N_c\rfloor, 2 \times N_p)$ in $VRSampling$.

As a next step (2), we need a method that splits D into $Dout$ and Din. This method should adhere to the fact that there exists a strong correlation between the out- and in-degrees of nodes. Additionally the sum of $Dout$ elements should be equal to the sum of Din, since it equals the number of graph arcs. A simple formula that adheres to both conditions is $\sum_{v \in V} D(v) = \sum_{v \in V} Dout(v) + \sum_{v \in V} Din(v)$.

Furthermore, we will randomly choose nodes (whose both out- and in-degree are bigger than 1) to assign them ($0.126 \times N_p$ in total) self-loops and pairs of nodes (whose both out- and in-degree are bigger than 2) to assign them ($0.177 \times N_p$ in total) multiple arcs (steps 3 and 4). Let S be the set of the chosen self-loops and M the set of the chosen multiple arcs. For each self-loop $\langle u, u \rangle$, we consider that $Dout(u) := Dout(u) - 1$ and $Din(u) := Din(u) - 1$. Similarly for each multiple arc $\langle u, v \rangle$, we consider that $Dout(u) := Dout(u) - 1$ and $Din(v) := Din(v) - 1$.

We can reduce the problem of generating a directed graph without self-loops and multiple-arcs given $Dout$ and Din as follows:

(LP1). Let G be a graph and let E be the set of its candidate arcs. The generation of G given its out- and in-degree sequences can be reduced to an LP instance of the form:

$$min\ \mathbf{0}^T \mathbf{x}$$
$$\sum_{(v,u) \in E} x_{v,u} = Dout(v),\ \forall v \in V$$
$$\sum_{(u,v) \in E} x_{u,v} = Din(v),\ \ \forall v \in V$$
$$0 \leq x_{u,v} \leq 1, \qquad \forall \langle u, v \rangle \in E$$

We consider as candidate edges all the $N \times (N-1)$ possible edges of a *directed graph* without *self-loops* and *multiple edges*. We should stress that, although we allowed the edges variables $x_{u,v}$ to take non-integer values, every solution of an *LP1* instance is integral (all the variables $x_{u,v} \in \{0,1\}$). This is ensured by the fact that (see Theorem 1) the matrix A of every *LP1* instance is (for a proof see [23]) *totally unimodular (TUM)*, i.e., every square submatrix of A has determinant equal to 0 or ± 1. It is well known [15,27] that if A is a TUM matrix, then all the vertices of the polytope $\{\mathbf{x} : A\mathbf{x} = \mathbf{b},\ \mathbf{x} \geq 0\}$ are integer for any integer vector $\mathbf{b}$. *Simplex* algorithm seeks for the optimum in the vertices of the polytope defined by an LP instance. Since, every vertex of the polytope defined by an *LP1* instance is integral every solution of it is integral (in our case $x_{u,v} \in \{0,1\}$).

Theorem 1. *Every LP1 instance can be described as $\{\mathbf{x} : A\mathbf{x} = \mathbf{b},\ 0 \leq \mathbf{x} \leq 1\}$ where A is a TUM $m \times n$ matrix.*

Attributes. In addition, we need to consider arcs which have as destination *Literals* (e.g., String, Integer), i.e., attributes of classes. After generating $\mathcal{G}_p$ we can add to the set of its nodes V, the Literal types, as specified in *XML schema*[2].

---

[2] http://www.w3.org/XML/Schema

Then we connect them to the pre-existent nodes of $\mathcal{G}_p$ under the condition that the total-degree sequence of $\mathcal{G}_p$ remains the same. This constraint can be satisfied by replacing a number k of arcs of the form $\langle u, v \rangle$, where u, v correspond to classes, such that $Dout(v) = 0$ (nodes representing literal types should have out-degree zero), with arcs of the form $\langle u, w \rangle$, where w is a node that corresponds to a *Literal* type. The number k of the attributes can be given as input (e.g., as a percentage of N_p).

Labeling properties. We should notice, that the output of $GenerateG_p$ is a bag of unlabeled edges, i.e. pairs of nodes. In the current implementation of PoweRGen, we consider a unique label (URI) per property in order to generate an RDFS schema. However, we provide the appropriate abstractions that are needed by a programmer to implement his one labeling policy. For instance, one might need the same property to have multiple domains / ranges. In that case, one should label with the same URI more than one edges.

3.3   Generating the Subsumption Graph

In order to generate $\mathcal{G}_s$ we consider as additional input the characteristic exponent, denoted by b, of the **power-law** of the PDF function of the *class descendants* distribution. Moreover, we take into account the depth, denoted by d, of $\mathcal{G}_s$ as well as the information whether $\mathcal{G}_s$ should be a DAG or a tree. Bellow, we detail $GenerateG_s$ algorithm.

Algorithm. $GenerateG_s$
Inputs.
b: the **PDF** exponent of the class descendants distribution, N_c: the number of schema classes.
(1) $Dout := PDFSampling(b, N_c - 1, 0.25 \times N_c)$;
(2) for $i = 1$ to $\lfloor 0.75 \times N_c \rfloor - 1$ do $D_{out} := D_{out} \uplus \{0\}$;
(3) $\gamma := 0.0017 \times N_c + 1.36$;
(4) $S := VRSampling(\gamma, d, N_c)$ (Order S in descending order);
(5) $k = \lfloor \frac{3}{4} \times d \rfloor$;
(6) $L :=$ the set of $\mathcal{G}_s^*$ levels ordered according to $f(l) = \|k - l\|$;

(7) for $i = 1$ to d do
       for $j = 1$ to $S[i]$ do
           $Din.append(L[i])$;

(8) Order $Dout$ in descending and Din in ascending order;

(9) $E :=$ the set of arcs of the solution of the *LP2*
      (or *LP3* if $\mathcal{G}_s$ is considered to be a DAG) instance
      corresponding to D_{in} and D_{out};

(10) if $\mathcal{G}_s$ is a Tree, then $E_s := E$;   else $E_s :=$ the transitive reduction of E;

Generating Dout of $\mathcal{G}_s^*$. Using the *PDFSampling* method, we can generate the *out-degree* sequence of $\mathcal{G}_s^*$. Specifically, the biggest allowed value is $N_c - 1$, since the *root* node has $N_c - 1$ descendants. Furthermore, on average the 75% of classes of real RDFS schemas are leaves, i.e. their out-degree is 0. Thus, we choose the following parameters $(b, M, N) = (b, N_c - 1, 0.25 \times N_c)$ for the sampling (step 1). To obtain a sequence of length N_c we add $0.75 \times N_c - 1$ times (corresponding to the 0.75% of leaf classes) the value 0 (step 2).

Generating Din of $\mathcal{G}_s^*$. The generation of Din of $\mathcal{G}_s^*$ is not as easy as that of $Dout$. This is due to the fact that no frequent pattern has been observed for the distribution of *class ancestors*. The best choice in this case, is to exploit the fact that the *classes per level* VR approximates a `power-law` for a significant proportion of real SW schemas (see Section 2.1). Moreover, the characteristic VR exponent, denoted by γ, approximately depends linearly on the number of nodes (classes). In this respect, we can produce a sequence of values, which correspond to the number of nodes that are located at a specific level. To this end, we sample according to the *classes per level* VR function with parameters $(b, N, sum) = (\gamma, d, N_c)$. This is due to the fact that we want to distribute N_c nodes to d levels (steps 3 and 4).

However, we still do not know to which level a specific value of the sampled set, denoted by S, corresponds. We exploit the fact that the level, denoted by k, at which the maximum number of nodes are located is approximately $0.75 \times depth$ for real SW schemas. Let x_i be the $i - th$ biggest value of S. We order levels of $\mathcal{G}_s^*$ according to their distance to the most populated level, i.e., $L = \langle k, k + 1, k - 1, k + 2, k - 2, ... \rangle$ (step 6). Then, we consider that x_i nodes are located at level $L(i)$.

Combining Dout and Din of $\mathcal{G}_p$. If $\mathcal{G}_s$ is a tree, the level at which a node is located coincides with the number of its transitive ancestors, i.e., its in-degree in $\mathcal{G}_s^*$ (step 7). However, if $\mathcal{G}_s$ is a DAG, the level of a node consists only the maximum lower bound for its in-degree. In the sequent, we use the symbol Din only for trees, while symbol Din^- for DAGs.

Trees We can reduce the problem of generating a tree given the degree sequences ($Dout$ and Din) of its TC as follows:

(LP2). Let $V_i = \{v \in V \mid Din(v) = i\}$ and d be the maximum value of i. V_i corresponds to the set of nodes located to the $i - th$ level of the tree G and d to its depth. Then the problem of generating G given the sequences of its transitive closure is reduced to the LP instance of the form:

$$min \ \mathbf{0}^T \boldsymbol{x}$$
$$\sum_{u \in V_{i+1}} (Dout(u) + 1) \times x_{v,u} = Dout(v), \ \forall i \in [0, d - 1], \ \forall v \in V_i$$
$$\sum_{u \in V_{i-1}} x_{u,v} = 1, \ \forall i \in [1, d], \ \forall v \in V_i$$
$$0 \leq x_{u,v} \leq 1, \ \forall \langle u, v \rangle \in E$$

To determine the candidate arcs (*LP2* variables) of G, we observe that, an arc $\langle u, v \rangle \in V \times V$ can be an arc of a Tree $G : (V, E)$ only if, $Dout(u) > Dout(v)$

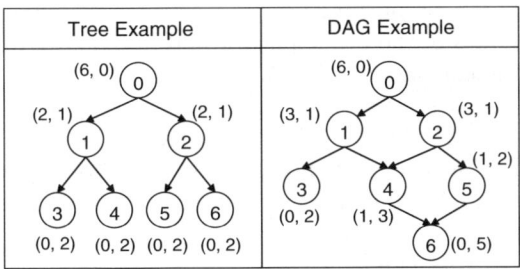

Fig. 3. An Example of a Tree (left) / DAG (right)

and $Din(u) = Din(v) - 1$. The first condition, (i.e., $Dout(u) > Dout(v)$) denotes that u should have more descendants than v and holds for every DAG (see for instance both parts of Figure 3), while the second one (i.e., $Din(u) = Din(v) - 1$) denotes that u should have one less ancestor than v and holds only for Trees. For instance, in the DAG of the right part of Figure 3, there exists an edge $\langle 2, 4 \rangle$, but $Din(2) = 1$ and $Din(4) = 3$, i.e., $Din(2) \neq Din(4) - 1$.

Moreover, it is evident that if arc $\langle u, v \rangle$ exists in the tree, then all v descendants, as well as v itself, are descendants of u. This fact is expressed by the constraints of the form $\sum_{u \in V_{i+1}} (Dout(u) + 1) \times x_{v,u} = Dout(v)$. The condition that every tree node has only one parent is captured by the constraints of the form $\sum_{u \in V_{i-1}} x_{u,v} = 1$.

We should notice that the value of every candidate edge is integral in every *LP2* instance solution (for a proof see [23]):

Theorem 2. *Every solution, $\boldsymbol{x}^*$ of an LP2 instance is integral. Specifically, it holds that $\boldsymbol{x}^* \in \{0, 1\}^n$.*

Directed Acyclic Graphs. We can reduce the problem of generating a TC DAG given *Dout* and *Din*⁻ as follows:

(LP3). Let G^* be a transitively closed DAG and E^* be the set of its candidate arcs. Then, the generation of G^* given its out-degree sequence, *Dout* and a sequence *Din*⁻, s.t., $Din^-(u)$ denotes the level of node u in G^*, can be reduced to an LP instance of the form:

$$min \ \boldsymbol{0}^T \boldsymbol{x}$$
$$\sum_{(u,v) \in E^*} x_{u,v} = Dout(u), \quad \forall u \in V$$
$$\sum_{(v,u) \in E^*} x_{v,u} \geq Din^-(u), \quad \forall u \in V$$
$$x_{u,v} + x_{v,w} - x_{u,w} \leq 1, \quad \forall \langle u, v \rangle, \ \langle v, w \rangle \in E^*$$
$$0 \leq x_{u,v} \leq 1, \quad \forall \langle u, v \rangle \in E^*$$

To determine the candidate arcs of G, we observe that, an arc $\langle u, v \rangle \in V \times V$ can be an arc of a transitively closed DAG $G^* : (V, E^*)$ only if, $Dout(u) > Dout(v)$ and $Din(u) < Din(v)$). We should notice that the strict inequalities guarantee that no cyclic path can be obtained by the considered set of candidate arcs[3]. Otherwise, all the nodes of the cyclic path would have exactly the same out-and in-degree, since G^* is transitively closed.

The constraint $x_{u,v} + x_{v,w} - x_{u,w} \leq 1$ guarantees that whenever both arcs $x_{u,v}, x_{v,w}$ exist (i.e., $x_{u,v} = x_{v,w} = 1$), the transitive arc $x_{u,w}$ also exists. This is justified as follows: whenever either $x_{u,v}$ or $x_{v,w}$ does not exist, the above inequality obviously holds, since it is of the form $a - b \leq 1$, where $a, b \in [0, 1]$. We focus on the case that both $x_{u,v}$ and $x_{v,w}$ exist. Then, $x_{u,w}$ also exists because G^* is transitively closed. Hence, $x_{u,v} + x_{v,w} - x_{u,w} = 1 + 1 - 1 \leq 1$, which obviously holds as equality.

We should mention that (unlike $LP1$ and $LP2$) the solutions of $LP3$ instances, are not integral in the general case. As a consequence, we should devise a method for obtaining the set of arcs of the generated graph. To this end, we can consider a threshold value $T \in [0, 1]$ and that an arc $\langle u, v \rangle$ exists iff $x_{u,v} \geq T$. In Section 4 we will examine threshold values that yield good approximations with respect to the given power-law.

Handling Infeasibility of LP instances. If the produced $LP2/LP3$ instance is infeasible, we conclude that there does not exist a tree/DAG whose transitive closure simultaneously realizes the given sequences. In that case we go back to step (1). *PDFSampling* uses a pseudo-random number generator and consequently a different *Dout* is produced for each *PDFSampling* method call.

3.4   Combining Both Graphs

At the last PoweRGen step we consider that $G_s : (V_s, E_s)$ and $G_p : (V_p, E_p)$ are generated. Since they have the same set of nodes we should define an one-to-one function $h : V_s \to V_p$ that maps each node of $\mathcal{G}_s$ to one and only node of $\mathcal{G}_p$. To this end, we exploit the fact that nodes with high out-degree in the $\mathcal{G}_p$ are located highly in the $\mathcal{G}_s$. Specifically, let k be the level of $\mathcal{G}_s^*$ at which the source nodes (corresponding to domains of properties) of most arcs of $\mathcal{G}_p$ are located. We order the nodes of V_p in descending out-degree order and we reach a list P. Also, let V_i be the set of nodes of $\mathcal{G}_s^*$ located at level i, i.e., $V_i = \{v \in V_s \mid Din(v) = i\}$. Then, we map each of the first $|V_k|$ nodes of P to one and only node of V_k. Similarly, we map the next $|V_{k+1}|$ nodes of P to one and only node of V_{k+1} and the next $|V_{k-1}|$ nodes of P to one and only node of V_{k-1}. This process continues until we map the nodes of V_0 and of V_d to nodes of V_p.

4   Experimental Evaluation

In this Section we experimentally evaluate PoweRGen on two axis, namely, the effectiveness and the efficiency of the proposed algorithm. Figure 4 (resp. Figure 5) shows the generated $\mathcal{G}_p$ (resp. $\mathcal{G}_s$) of a schema with 300 classes, 1000

---

[3] Replace inequalities with equalities to allow cyclic paths.

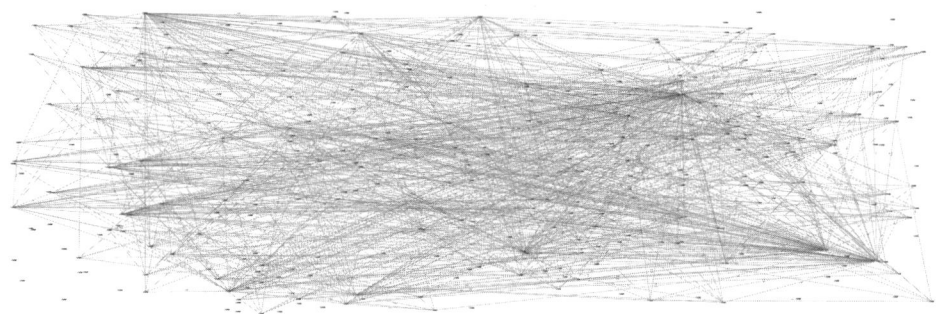

Fig. 4. Synthetic $\mathcal{G}_p$ example

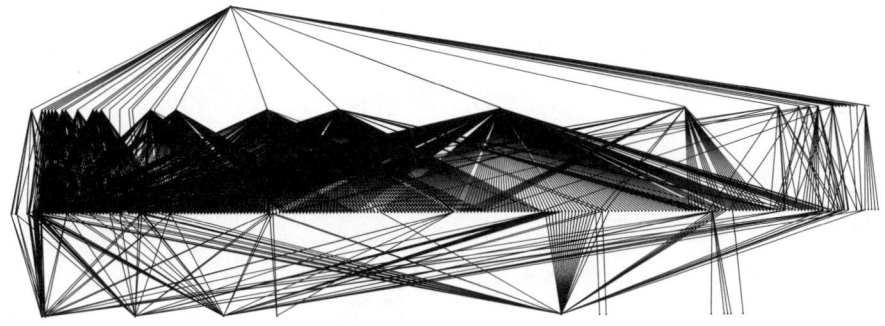

Fig. 5. Synthetic $\mathcal{G}_s$ example

properties, total-degree VR exp. 0.6, *depth* = 5 and class descendants PDF exp. 1.7. The figures have been drawn using *StarLion* [26] and *VRP* [1] visualization facilities.

4.1 Effectiveness

To investigate to what extent a function approximates a `power-law`, we rely on a commonly used method (based on the least square errors method), called Linear Regression [20], to fit a line in a set of 2-dimensional points and, thus, to investigate whether the *log-log* plot of a function approximates a line. The accuracy of the approximation is indicated by the correlation coefficient, the absolute value of which (hereafter called *ACC*) always lies in [0, 1]. An *ACC* value 1.0 indicates perfect linear correlation, i.e., the points are exactly on a line. Additionally, if PDF follows a `power-law` with exponent β, i.e., $P(X = x) = \alpha x^{-\beta}$, then CCDF follows a `power-law` with exponent $\beta - 1$, i.e., $P(X \geq x) = \gamma x^{-(\beta-1)}$ (see [7] for details). Based on this fact, we do *Linear Regression* on CCDF plots [7] and the results regarding PDF immediately follow.

Total degree CCDF and VR functions of $\mathcal{G}_p$ are illustrated in Figure 6 (upper). As one can observe the CCDF is almost a `power-law` ($ACC = 0.98$). This small

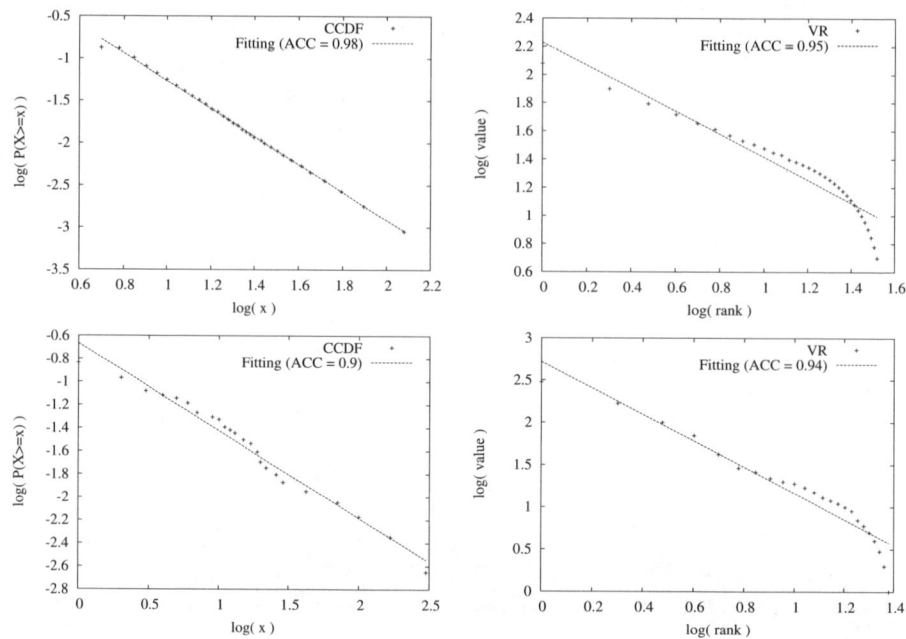

Fig. 6. $\mathcal{G}_p$ total-degree and $\mathcal{G}_s^*$ out-degree functions of our example

divergence from a strict **power-law** (i.e., $ACC = 1.0$) is due to the *VRSampling* method (see Section 2.1).

In the case of $\mathcal{G}_p$ and trees for $\mathcal{G}_s$, we generate schemas whose degree sequences are sampled by the **power-law** distributions assumed as input. On the other hand, in the case of DAGs for $\mathcal{G}_s$ the *LP3* instance solutions usually have coordinates (corresponding to arcs of $\mathcal{G}_s^*$) that are real numbers in $(0, 1)$. To reach a final set of arcs we will consider a threshold value $T \in [0, 1]$: an arc $\langle u, v \rangle$ exists iff $x_{u,v} \geq T$. The use of such heuristic may miss some transitive arcs of $\mathcal{G}_s^*$. For instance, consider that $T = 0.6$ and in the LP solution $x_{u,v} = 0.7$, $x_{v,w} = 0.8$ and $x_{u,w} = 0.5$. Then, $x_{u,v} + x_{v,w} - x_{u,w} = 1 \leq 1$, as required for the arcs of a TC DAG. However, to make the solution integral, we consider that $x_{u,v} = x_{v,w} = 1$ and $x_{u,w} = 0$. Concerning the value of T, we should notice that a big threshold value (e.g., $T = 0.9$), decreases the probability of missing transitive arcs, but ignores many arcs and thus the generated *Dout* and *Din* roughly diverge from the given ones. On the other hand, a small threshold value (e.g., $T = 0.1$), results in bigger number of arcs than in the given sequences. After experiments we found that $T = 0.6$ leads to best approximations with respect to the given **power-law**.

To measure the quality of the approximations yielded by the above threshold, we generated 1000 subsumption DAGs with parameters: 300 classes, *depth* = 5 and $b = 1.7$. Table 1 shows the distribution of the exponents and the *ACC* values of the generated subsumption graphs. The average computed value for b is

Table 1. PDF `power-law` exponent and ACC of 1000 subsumption graphs with given $b = 1.7$

	Min	Max	Mean	St.dev.	COV
b	1.6	1.767	1.691	0.032	0.047
ACC	0.874	0.922	0.903	0.006	0.007

1.691, while the average ACC value is 0.903. We conclude that the out-degree distribution of $\mathcal{G}_s^*$ approximates ($ACC = 0.903$, i.e., 90.3%) a `power-law` whose characteristic PDF exponent approximates ($\frac{1.691}{1.7} = 0.994$, i.e., 99.4%) 1.7. Figure 6 (bottom) illustrates the CCDF and the VR functions of the out-degree of one of the 1000 generated subsumption graphs.

4.2 Efficiency

In this Section we report the time and memory requirements of the LP instances produced for the generation of synthetic SW schemas of various size. To solve the LP instances we used the academic software Soplex [29]. The measurements regarding memory and runtime reported below are strongly depended on the efficiency of Soplex. Improved measurements can be obtained by using more sophisticated *simplex* implementations, such as the commercial CPLEX [16]. We also measure the number of variables and constraints of the produced LP instances that show the complexity of the reduction of our problem to the LP instances and are independent of the specific software implementation of *simplex*. Experiments were carried out on a PC with a Pentium IV 3.2GHz processor and 2 GB of main memory, over Suse Linux (v10.1).

Figure 7 (upper left) shows the number of variables and constraints of the LP1 instances that are produced to generate $\mathcal{G}_p$ for various number of classes, namely $300-1000$. Specifically, we considered as constant $b = 0.6$ and $N_p = 1000$ and we vary N_c. Figure 7 (upper right) shows the time and space requirements of Soplex to solve the *LP1* instances produced to generate $\mathcal{G}_p$. As one can observe, the generation of typically sized schemas with 700 classes and 1000 properties [25] needs 60 sec and 114 MB memory.

Figure 7 (middle left) shows the number of variables and constraints of the *LP3* instances that are produced to generate DAGs for $\mathcal{G}_s$ for various number of classes, namely $100-700$. Specifically, we considered as constant $b = 2.2$, $depth = 7$ and vary N_c. Figure 7 (middle right) shows the time and space requirements of Soplex to solve the *LP3* instances. We should mention that Soplex crashes for schemas with more than 700 classes. This is due to the transitivity constraints, whose number equals the number of all triples u, v, w of nodes, such that $\langle u, v \rangle$, $\langle v, w \rangle$ are candidate arcs.

On the other hand, in case of trees, the generation of $\mathcal{G}_s$ is much more efficient. To show this fact we considered the degree sequences of the *transitive closures* of trees whose 75% of nodes are leaves and we reduce their generation to an *LP2* instance. We exponentially increase the number of classes, which is computed as a

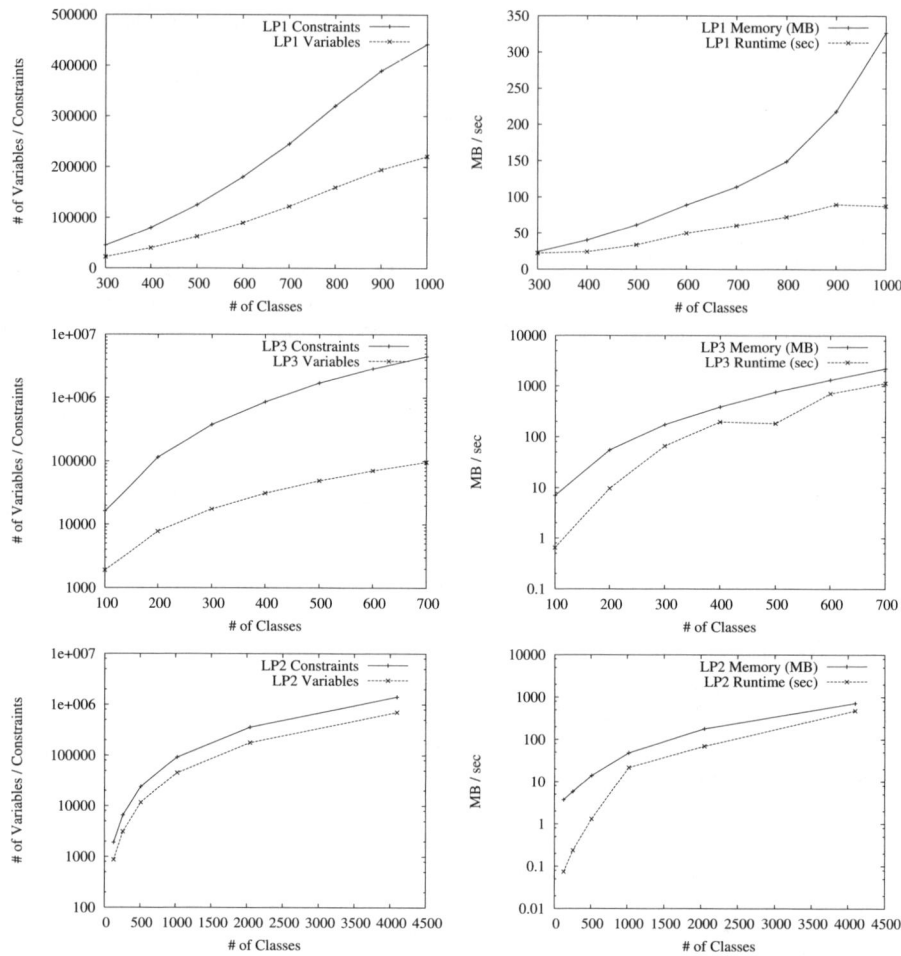

Fig. 7. Efficiency measures for *LP1* (upper) / *LP3* (middle) / *LP2* (bottom) instances

power of 2. Figure 7 (bottom left) shows the number of variables and constraints of the produced *LP2* instances, while Figure 7 (bottom right) the corresponding runtime and memory requirements.

4.3   Experimental Conclusions

As a general remark, PoweRGen produces synthetic property and subsumption graphs whose distributions respect the `power-law` exponents given as input with a confidence ranging between $90 - 98\%$. Regarding efficiency, PoweRGen is scalable concerning the generation of G_p and of G_s in case of trees. On the other hand, in the case of DAGs for G_s, the memory needs rapidly increase with

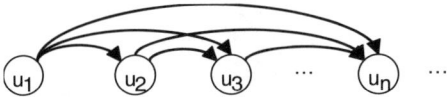

Fig. 8. Algorithm [12,13] Generating Undirected Graphs

respect to the number of classes. However, we should notice, that, even in that case, we can generate schemas up to 700 classes, which exceeds the class number of most real RDFS schemas [23].

5 Related Work

To the best of our knowledge, there does not exist any related work on a general purpose synthetic RDFS schema generator. A recent work [22] has proposed a method to generate small sized schemas by varying the URIs of resources. The motivation of that generator was to evaluate ontology mapping methods. Therefore the focus on the generation process was given to the way that URIs are produced, e.g. a URI is substring of an other. On the other hand, PoweRGen imitates graph features RDFS schemas frequently exhibit in reality and the URIs of the produced properties are of very minor interest, e.g. we assign typical URIs of the form Prop1, Prop2 etc. [11,19] propose a simplistic method to generate SW data as instances of an RDFS (or OWL) schema given as input.

Moreover, [21,2] focus on the generation of XML data that are valid against either a predefined DTD [21] or an XML schema given by the user [2]. However, the full potential of the SW lies on the existence of schemas, which can be exploited to support advanced reasoning services against the available SW data. As a consequence, we need to also generate the schemas, except for the data. Note also, that SW schemas (as well as instance descriptions) are actually graphs, while XML documents are trees.

In addition, [12,13] considered only the generation of simple graphs (undirected, without self-loops and multiple-arcs). Their work is based on the result of [8,10], which can be summarized as follows. Let n denote the number of nodes of the graph we wish to generate. Let u_i, $1 \le i \le n$ denote nodes and $d_1 \ge d_2 \ge \ldots \ge d_n$ intended degrees of these nodes. The necessary and sufficient condition for a degree sequence to be realizable is: $\sum_{i=1}^{k} d_i \le k(k-1) + \sum_{i=k+1}^{n} min\{k, d_i\}$.

The algorithm is iterative and maintains the residual degrees of vertices. In each iteration step it picks an arbitrary vertex u and adds arcs from u to d_u vertices of highest residual degree, where d_u is the residual degree of u (see Figure 8). The residual degrees of the latter d_u vertices are updated appropriately. By connecting with d_u highest degree vertices the algorithm ensures that the necessary and sufficient condition holds for the residual problem instance. This algorithm is not suitable for synthetic SW schema generation. Specifically, for the generation of G_s we need to consider the arcs transitivity. Moreover, both G_s and G_p are directed graphs.

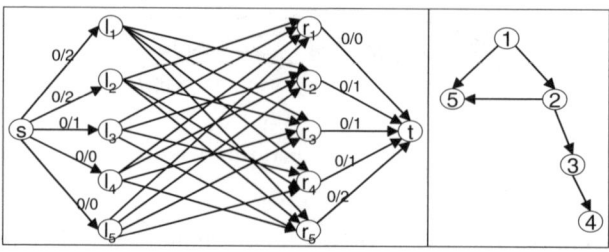

Fig. 9. Algorithm [18] Generating Directed Graphs (left) and its output (right)

Moreover, [18] proposed a reduction of the problem of generating *directed* graphs that simultaneously satisfy two given degree sequences to the *Max Flow* problem. In particular, let $d_{in} = (d_{in,1}, d_{in,2}, ..., d_{in,n})$ and $d_{out} = (d_{out,1}, d_{out,2}, ..., d_{out,n})$ be sequences of integers (in no particular sorted order), with $\sum_{i=1}^{n} d_{in,i} = \sum_{i=1}^{n} d_{out,i}$. We wish to construct a directed graph on n nodes, such that node u_i has $d_{in,i}$ incoming arcs and $d_{out,i}$ outgoing arcs, $1 \le i \le n$. Consider the graph that has a source s, a sink t, a set of nodes $L = \{l_1, ..., l_n\}$ and a set of nodes $R = \{r_1, ..., r_n\}$. There is a link of capacity 1 directed from each l_i to each r_j, for $1 \le i, j \le n$ and $i \ne j$. There is a link of capacity $d_{out,i}$ directed from s to each l_i, for $1 \le i, j \le n$. Finally, there is a link of capacity d_{in}, i directed from each r_i to t, for $1 \le i, j \le n$. We may now consider integral maximum flows from s to t. If there is such a flow of value $\sum_{i=1}^{n} d_{in,i} = \sum_{i=1}^{n} d_{out,i}$, then the corresponding degree sequences are simultaneously realizable and the flow gives a directed graph that satisfies, simultaneously, in-degrees $d_{in,i}$ and out-degrees $d_{out,i}$.

Figure 9 illustrates an example for this algorithm, where $d_{in} = \langle 0, 1, 1, 1, 2 \rangle$ and $d_{out} = \langle 2, 2, 1, 0, 0 \rangle$. The maximum integral flow of the graph of Figure 9 (left), is $maxFlow = 5 = \sum_{d \in d_{in}} d = \sum_{d \in d_{out}} d$. Hence, d_{in}, d_{out} are simultaneously realizable and the graph of the maximum flow, which is drawn in Figure 9 (right), satisfies them. This algorithm can be used for the generation of G_p, but not of that of G_s since: a) the arcs transitivity needs to be considered and b) G_s is a DAG (i.e., no cycles are allowed).

Furthermore, [5] studied the number of *simple undirected graphs* that realize a given sequence and proposed non deterministic algorithms to generate one of them. Specifically, they focused on how each one of the graphs that realize the given sequence is generated from their algorithms with as uniform distribution as possible. Although theoretically interesting, that work cannot be used in our case, given that it considers undirected graphs (while both G_p and G_s are directed) and also ignores the transitivity of arcs (needed to be considered for G_s).

Finally, there exist works that generate graphs exhibiting `power-law` degree distributions (see [7] for a survey). However, a) the `power-law` arise after a big number of node insertions in the graph, b) the characteristic exponent is fixed or is constrained to big values (e.g., bigger than 2) and c) the transitivity of arcs is not handled. Hence, we can not use these generators, because a) RDFS schema graphs are relatively small sized, b) the `power-law` exponent should be fully parameterizable (it lies in $[0.79, 2.18]$ for total-degree VR, while in $[1.54, 2.47]$ for class descendants PDF for real schemas) and c) we should enforce as much as possible arcs transitivity for $\mathcal{G}_s$.

6   Future Work

As future work, we plan to explore the possibility of devising a more efficient method to generate $\mathcal{G}_s$ when it is considered to be a DAG. To this end, it would be interesting to investigate if graph generation methods exploiting the ideas of *self-similarity* and *fractals* [17], could be modified to take as input the degree sequences, or at least the characteristic `power-law` exponent of their distribution. We also plan to exploit the algorithms presented in this paper for synthetic SW data generation as a part of our ongoing *RDF/S* benchmarking efforts [24].

References

1. Alexaki, S., Christophides, V., Karvounarakis, G., Plexousakis, D., Tolle, K.: The ICS-FORTH RDFSuite: Managing Voluminous RDF Description Bases. In: 2nd International Workshop on the Semantic Web (May 2001)
2. Barbosa, D., Mendelzon, A., Keenleyside, J., Lyons, K.: ToXgene: A template-based data generator for XML. In: Proceedings of the ACM SIGMOD International Conference on Management of Data, New York, p. 616 (2002)
3. Barbosa, D., Mignet, L., Veltri, P.: Studying the XML Web: Gathering Statistics from an XML Sample. World Wide Web Journal 8(4) (2005)
4. Berners-Lee, T., Hendler, J., Lassila, O.: The Semantic Web. Scientific American (May 2001)
5. Blitzstein, J., Diaconis, P.: A Sequential Importance Sampling Algorithm For Generating Random Graphs With Prescribed Degrees. Annals of Applied Probabilty (2005)
6. Brickley, D., Guha, R.V.: RDF Vocabulary Description Language 1.0: RDF Schema, W3C Recommendation (February 10, 2004)
7. Chakrabarti, D., Faloutsos, C.: Graph Mining: Laws, Generators, and Algorithms. ACM Computing Surveys (CSUR) 38(2) (2006)
8. Claude, B.: Graphs and Hypergraphs. North Holland Publishing Company, Amsterdam (1973)
9. Wang, T.D.: Gauging Ontologies and Schemas by Numbers. In: Proc. Fourth Int'l Workshop Evaluation of Ontology-Based Tools (EON) (2006)
10. Erdös, P., Gallai, T.: Graphs with Prescribed Degree of Vertices. Mat. Lapok 11, 264–274 (1960)
11. Guo, Y., Heflin, J., Pan, Z.: Benchmarking DAML+OIL Repositories. In: Fensel, D., Sycara, K.P., Mylopoulos, J. (eds.) ISWC 2003. LNCS, vol. 2870, pp. 613–627. Springer, Heidelberg (2003)
12. Hakimi, S.: On the Realizability of a Set of Integers as Degrees of the Vertices of a Graph. SIAM 10, 496–506 (1962)

13. Havel, V.: A Remark on the Existence of Finite Graphs. Casopis Pest. Mat. 80, 477–480 (1955)
14. Hayes, P.: RDF Semantics. W3C Recommendation (February 10, 2004)
15. Hoffman, A.J., Kruskal, J.B.: Integral Bounding Points of Convex Polyedra. In: Linear Inequalities and Related Systems, pp. 223–246. Princeton University Press, Princeton (1956)
16. Hsu, M., Cheatham, T.E.: Rule Execution in CPLEX: A Persistent Objectbase. In: Dittrich, K.R. (ed.) OODBS 1988. LNCS, vol. 334, pp. 150–155. Springer, Heidelberg (1988)
17. Leskovec, J., Chakrabarti, D., Kleinberg, J.M., Faloutsos, C.: Realistic, Mathematically Tractable Graph Generation and Evolution, Using Kronecker Multiplication. In: Jorge, A.M., Torgo, L., Brazdil, P.B., Camacho, R., Gama, J. (eds.) PKDD 2005. LNCS (LNAI), vol. 3721, pp. 133–145. Springer, Heidelberg (2005)
18. Mihail, M., Visnoi, N.: On Generating Graphs with Prescribed Degree Sequences for Complex Network Modeling Applications. In: Procs of Approx. and Randomized Algorithms for Communication Networks (ARACNE) (2002)
19. Perry, M.: Test ontology generation tool,
    `http://lsdis.cs.uga.edu/projects/semdis/tontogen/`
20. Press, W.H., Teukolsky, S.A., Vetterling, W.T., Flannery, B.P.: Numerical Recipes in C, 2nd edn. Cambridge University Press, Cambridge (1992)
21. Schmidt, A., Waas, F., Kersten, M., Carey, M.J., Manolescu, I., Busse, R.: Xmark: A benchmark for XML data management. In: Procs of the 28th International Conference on Very Large Data Bases, Hong Kong, China, pp. 974–985 (2002)
22. Svab, O., Svatek, V.: Vitro Study of Mapping Method Interactions in a Name Pattern Landscape. In: 2nd International Workshop on Ontology Matching, collocated with ISWC 2007 (2007)
23. Theoharis, Y.: On Power Laws and the Semantic Web. Master's thesis, Computer Science Department, University of Crete (February 2007),
    `http://athena.ics.forth.gr:9090/RDF/publications/`
    `MasterThesisTheohari.pdf`
24. Theoharis, Y., Christophides, V., Karvounarakis, G.: Benchmarking Database Representations of RDF/S Stores. In: Gil, Y., Motta, E., Benjamins, V.R., Musen, M.A. (eds.) ISWC 2005. LNCS, vol. 3729, pp. 685–701. Springer, Heidelberg (2005)
25. Theoharis, Y., Tzitzikas, Y., Kotzinos, D., Christophides, V.: On Graph Features of Semantic Web Schemas. IEEE Transactions on Knowledge and Data Engineering 20(5), 692–702 (2008)
26. Tzitzikas, Y., Kotzinos, D., Theoharis, Y.: On Ranking RDF Schema Elements (and its Application in Visualization). Journal of Universal Computer Science, Special Issue: Ontologies and their Application 13(12), 1854–1880 (2007)
27. Veinott, A.F., Dantzig, G.B.: Integral Extreme Points. SIAM Review 10(3), 371–372 (1968)
28. W3C. W3C Semantic Web Activity, W3C Workshop on RDF Access to Relational Databases (October 25-26, 2007)
29. Wunderling, R.: Paralleler und Objektorientierter Simplex-Algorithmus. Ph.D. thesis, ZIB (1996)

Ontology-Based Data Sharing in P2P Databases

Dimitrios Skoutas[1,2], Verena Kantere[1], Alkis Simitsis[3,*], and Timos Sellis[1,2]

[1] School of Electrical and Computer Engineering
National Technical University of Athens, Athens, Hellas
{dskoutas,verena}@dblab.ece.ntua.gr
[2] Institute for the Management of Information Systems (R.C. "Athena")
Athens, Hellas
timos@imis.athena-innovation.gr
[3] Stanford University
Palo Alto, California, USA
alkis@db.stanford.edu

Abstract. We consider peer-to-peer systems in which peers share structured data through the use of schema mappings. Peers express their queries and rewrite incoming queries on their local schema. We assume the existence of one or more ontologies describing the domain of interest of the peers. The ontologies are used to semantically annotate each peer schema, making explicit the type of information provided by it. A major problem in such a system is that peers cannot easily judge the semantic relativeness of their interests to interests of other peers, as these are expressed by the respective local schemas. Moreover, peers cannot evaluate the semantic relativeness of answers that they receive to their queries. In this paper, we propose a semantic similarity measure for evaluating the semantic relativeness between peer schemas, as well as between queries and their rewritten versions on other peers. The similarity measure is first introduced under the assumption of a shared ontology among the community of peers, and then it is extended, employing ontology matching and translation techniques, to support the comparison of class expressions accross multiple ontologies. The proposed similarity measure adopts the notions of recall and precision from the field of Information Retrieval. Our goal is to use this measure for the identification of semantically relevant peers and the evaluation of the quality of the received answers based on the semantic annotations, the mappings, and the queries issued.

1 Introduction

Peer-to-peer overlays (hereafter P2P) have been consistently used in the previous years for massive sharing and exchange of unstructured data. Emerging applications of the P2P paradigm are the Peer Data Management Systems (PDMS's) -e.g., [1,2]- which hold a leading role in sharing semantically rich information. PDMS's consist of autonomous sources that store and manage structured data locally, revealing part of their local data schema to the rest of the peers. Pure P2P systems -i.e., without super-peers- are considered to operate in lack of a global schema. Without a reference schema, peer

---

* This work was performed while the author was with IBM at IBM Almaden Research Center, San Jose, California.

V. Christophides et al. (Eds.): SWDB-ODBIS 2007, LNCS 5005, pp. 117–137, 2008.
© Springer-Verlag Berlin Heidelberg 2008

databases express and answer queries based on their local schema. In particular, peers that are directly linked, i.e., *acquainteed*, establish a common way of exchanging and comprehending each others' data. Usually this is realized in the form of mappings between the peer schemas. Using the peer mappings and some suitable rewriting algorithm, two acquainted peers can propagate queries to each other.

The nature of structured data stored in the overlay enforces strict methods of querying and query rewriting. However, frequently, the user intends to obtain information that is semantically relevant to the posed query, rather than information that strictly complies to structural constraints. The available rewriting algorithms for structured data target the classic data integration problem [3] and consider only queries that can be completely rewritten to the target schema under a set of mappings. Still, such approach is not enough for a P2P environment where peers seek and are satisfied with information semantically similar, but not necessary identical, to their requests (as in the case of popular P2P file sharing applications).

An example application where the semantic similarity plays a significant role, is the creation of *social networks*. Recently, new social networking services have been emerging, that are similar to human social networks. Services such as MySpace[1] and Orkut[2], to mention a few, form virtual communities, with each participant setting his/her own characteristics and interests. Their goal is to allow members to form relationships through communication with other members and sharing of common interests. In these applications, the search for identical information among the users is not realistic.

Hence, there is a necessity for investigating the notion of semantic similarity of peer schemas, and furthermore peer queries, with their rewritten versions. Using such similarity criteria, users can identify peers sharing similar interests to theirs. For each specific query they pose, the system can decide which peers can rewrite it better and, thus, give more satisfying answers. Peer schemas and query rewritings can be ranked according to their semantic relativeness to a reference schema or to an original query, respectively.

Nevertheless, it is not straightforward to encounter the semantic similarity problem in the context of structured data without any additional semantic information [4]. Database schemas and respective mappings cannot capture sufficient semantic metadata so that a qualitative solution for the semantic similarity problem can be anticipated. In our work, to deal with this problem, we rely on Semantic Web technology. Specifically, we consider a PDMS accompanied by one or more domain ontologies, which are used to semantically annotate the content a peer makes available to the network. Note that the use of these ontologies does not contravene the requirements of the lack of global schema and peer autonomy: peer schemas do not have to adhere to any restrictions; they may just use terms from these ontologies to semantically describe their elements.

Contributions. Our main contributions are as follows.

- We address the problem of semantic similarity of schemas and queries for a PDMS enhanced with one or more domain ontologies.

---

[1] http://www.myspace.com

[2] http://www.orkut.com

- We propose the use of the measures *recall* and *precision* for quantifying the notion of semantic similarity.
- We propose a combined similarity measure, taking into consideration (a) the semantics of the peers' schemas, (b) the mappings between the peers, and (c) the queries issued by the peers.
- We extend the proposed similarity measure for the comparison of elements across different ontologies.

Outline. In the rest of this paper, Section 2 formally defines the problem under consideration and introduces a running example used throughout the paper to clarify the concepts introduced. Section 3 shows how the measures *recall* and *precision* can be used for the semantic comparison of peer schemas. Section 4 describes how the techniques may extend to queries and mappings. Section 5 extends the proposed similarity measures to deal with semantic annotations derived from different ontologies. Finally, Section 6 demonstrates the state of the art, while Section 7 concludes the paper with a prospect of the future.

2   Framework and Problem Description

2.1   Preliminaries

The Semantic Web [5] is emerging as a vision to enhance the current Web with machine-processable metadata, specifying the intended meaning of the provided information in a formal and explicit manner. Software agents can then leverage these metadata to "understand", process, and reason about the described resources, therefore increasing the degree of automation, the efficiency, and the effectiveness of searching, sharing, and combining information. Ontologies have a central role in this effort. An ontology can be defined as a formal specification of a shared conceptualization [6]. OWL has been proposed by W3C as a recommendation for a language for specifying ontologies on the Web [7]. OWL is based on Description Logics [8], a decidable fragment of first-order logic, constituting an important and commonly used knowledge representation formalism. Using OWL, one can describe the knowledge about a particular domain in terms of

- a set of *classes*, representing the entities of interest in the domain of discourse, and
- a set of *properties*, representing attributes of these entities or relationships between them.

In particular, two types of properties are provided:

- *object properties*, which relate instances of one class to instances of another class, and
- *datatype properties*, which relate instances of one class to values of a specified datatype.

Classes, as well as properties, may be organized in an appropriate *hierarchy*. Furthermore, it is possible to specify *restrictions* on the values and/or the minimum and/or

OWL construct	Notation	Description
owl:Class	C	Classes
owl:ObjectProperty	P	Object properties
owl:DatatypeProperty	P	Datatype properties
owl:equivalentClass	$C_1 \equiv C_2$	Class equivalence
rdfs:subClassOf	$C_1 \sqsubseteq C_2$	Class subsumption
owl:equivalentProperty	$P_1 \equiv P_2$	Property equivalence
rdfs:subPropertyOf	$P_1 \sqsubseteq P_2$	Property subsumption
owl:Thing	$\top$	The class containing all the individuals
owl:Nothing	$\bot$	The class containing no individuals
owl:DataRange	d	Data types
rdfs:domain	$domain(P)$	The domain of a property
rdfs:range	$range(P)$	The range of a property
owl:allValuesFrom	$\forall P.C$	Value restrictions on object properties
owl:allValuesFrom	$\forall P.d$	Value restrictions on datatype properties
owl:minCardinality	$\geq_n P$	Min cardinality restriction
owl:maxCardinality	$\leq_n P$	Max cardinality restriction

Fig. 1. OWL constructs and notation used

maximum cardinality of a property with respect to a specific class. Finally, the use of custom data types can be provided by an extension of OWL, such as OWL 1.1 [9] or OWL-Eu [10]. Figure 1 summarizes the OWL constructs and notation used throughout the paper. Additionally, we use the notation $P(C)$ to denote the set of properties that are related to a class C, and the notation $R(P_C)$ to denote the set of restrictions on a property P with respect to a class C.

2.2   Problem Description

We consider an unstructured PDMS. Each peer possesses a local database exposing a relational schema. Queries are issued according to this schema. Each peer shares data with its acquainted peers via a set of mappings, which are used for query rewriting between the respective schemas. We focus our study on SPJ queries, and mappings of the well known forms GAV-LAV-GLAV [3] as they are adapted to the P2P paradigm [11]. A query Q specifies the information to be retrieved, by means of a set of attributes to be returned (SELECT clause), and a set of conditions to be applied (WHERE clause).

In addition, we assume the existence of a domain ontology, providing a shared conceptualization of the domain of interest for the community of peers (observe Figure 2a.) The domain ontology may be provided by a third-party, such as a standardization organization. This is a realistic hypothesis for a wide range of applications, involving networks where peers are professionals, companies, or organizations (e.g., universities, libraries, hospitals), exchanging information about a specific domain. An alternative case is the collaborative construction of the ontology within the peer network itself. In large-scale peer-to-peer networks, where global consensus is difficult to achieve and

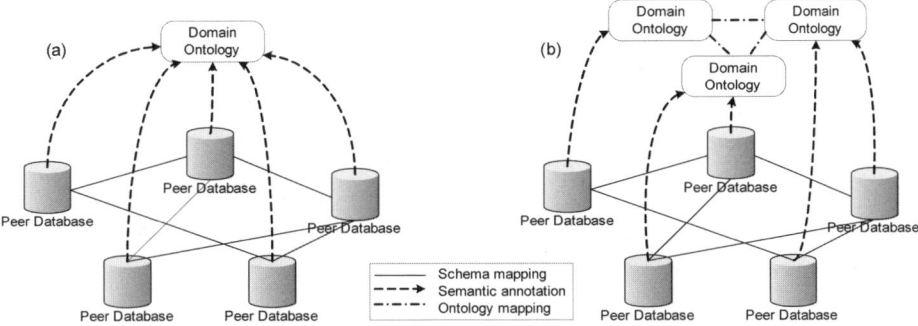

Fig. 2. Unstructured network of semantic peers with (a) single ontology (b) multiple ontologies

maintain, there may exist several ontologies, allowing different views of the domain, and clusters of peers using either of these ontologies (observe Figure 2b.) In these cases the proposed approach is still applicable, provided that mappings between these ontologies are available. For simplicity, we first introduce the proposed similarity measure assuming the existence of a shared ontology (Sections 3 and 4), and then we extend it to deal with the case of multiple ontologies (Section 5).

The domain ontology is used to semantically annotate the schemas of the peers participating in the network, describing the type of information that a peer makes available to other peers. The semantic annotation of a peer's schema is achieved by declaring correspondences between terms in the peer schema and terms in the domain ontology. The high-level architecture of the system is depicted in Figure 2. Solid lines represent pairwise mappings between acquainted peers, while dashed lines represent correspondences between elements of the local schema and elements of the domain ontology.

Definition 1. *A semantic peer is a tuple* $\mathcal{P} = (\mathcal{R}, \mathcal{O}, \mathcal{A})$, *where* $\mathcal{R}$ *is the peer's database schema,* $\mathcal{O}$ *the domain ontology used, and* $\mathcal{A}$ *the peer's semantic annotation.* $\mathcal{A}$ *holds the set of annotations A for the relations in* $\mathcal{R}$. *Each A consists of a pair of the form (R,C) and a set of pairs of the form* $(R.t, C_i.P)$, $C_i \in \mathcal{C}$. *That is, a relation R is semantically annotated by means of a set of classes C. Each* $C_i \in \mathcal{C}$ *is an ontology class, possibly enhanced with additional restrictions to make explicit the semantics of the underlying relation, i.e.,* $C_i \equiv C_i' \sqcap_k R_k$, $C_i' \in \mathcal{O}$. *Attributes R.t are annotated by means of properties of the same set of classes, i.e.,* $(R.t, C_i.P)$, $C_i \in \mathcal{C}$.

Our work aims at the exploitation of the information conveyed by the ontology and the annotations to provide a measure that represents how semantically close are two peers $\mathcal{P}_1$ and $\mathcal{P}_2$, namely $Sem\_Sim(\mathcal{P}_1, \mathcal{P}_2)$. Furthermore, when a query Q is forwarded by $\mathcal{P}_1$ to $\mathcal{P}_2$ and is rewritten to Q' based on the corresponding mappings, then it is usually degraded. This means that some part of Q cannot be rewritten on peer $\mathcal{P}_2$ [4]. Hence, our goal is to provide a measure of the degree of match between the requested information Q and the retrieved information Q', i.e., $Sem\_Sim(\mathcal{P}_1, \mathcal{P}_2, Q, Q')$.

Having such qualitative measures for the semantic relationship between the peers is important, when, for instance, a peer chooses its acquaintances or evaluates the quality of the answers returned by another peer.

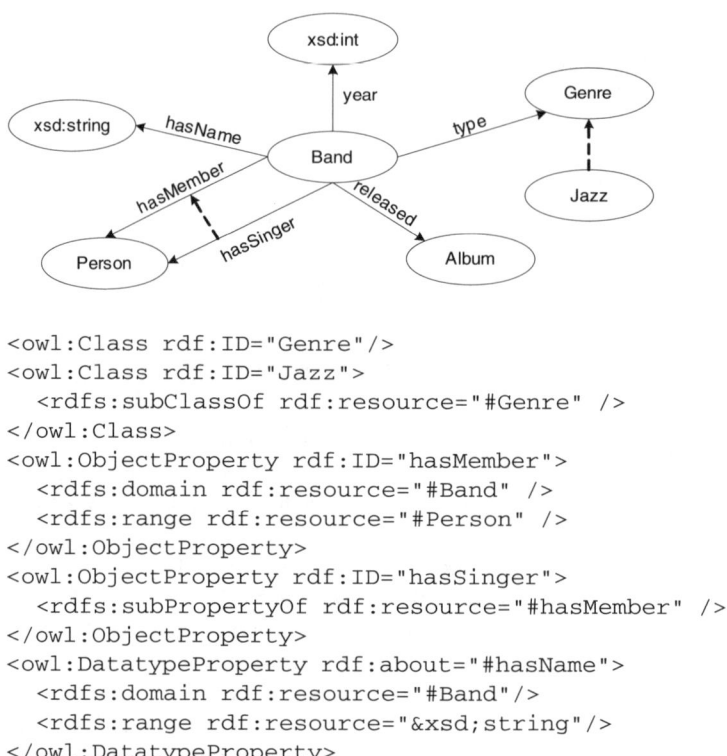

```
<owl:Class rdf:ID="Genre"/>
<owl:Class rdf:ID="Jazz">
   <rdfs:subClassOf rdf:resource="#Genre" />
</owl:Class>
<owl:ObjectProperty rdf:ID="hasMember">
   <rdfs:domain rdf:resource="#Band" />
   <rdfs:range rdf:resource="#Person" />
</owl:ObjectProperty>
<owl:ObjectProperty rdf:ID="hasSinger">
   <rdfs:subPropertyOf rdf:resource="#hasMember" />
</owl:ObjectProperty>
<owl:DatatypeProperty rdf:about="#hasName">
   <rdfs:domain rdf:resource="#Band"/>
   <rdfs:range rdf:resource="&xsd;string"/>
</owl:DatatypeProperty>
```

Fig. 3. A sample ontology and a snippet of the corresponding XML representation

2.3 Motivating Example

As a motivating example, consider a simple scenario where in the context of a social
network system, two peers, P_1 and P_2, want to exchange data about music bands. Suppose that P_1 and P_2 have the following schemas and mapping:

$$P_1 : bands(name, members, year)$$
$$P_2 : bands(name, singer, year) \tag{1}$$
$$M_{P_1,P_2} : bands(name, members, year) :- bands(name, singer, year)$$

An ontology for the music domain is used to describe semantically the contents of
the peers. A sample snippet of such ontology is illustrated in Figure 3. Nodes represent
classes or datatypes, while edges represent properties. Dashed lines between two classes
or properties represent subsumption relation.

Furthermore, we assume that P_1 contains information about bands having at least
one album, while P_2, being more specialized, stores information about bands that play
Jazz music, were formed before the year 2000, and have released at least 3 albums.
These facts can be made explicit by each peer, by annotating the relations and attributes

Schema element	Ontology element
bands.name	hasName
bands.members	hasMember
bands.year	year
bands	$Band\_P_1$

Schema element	Ontology element
bands.name	hasName
bands.singer	hasSinger
bands.year	year
bands	$Band\_P_2$

(a) Peer $\mathcal{P}_1$ (b) Peer $\mathcal{P}_2$

Fig. 4. Semantic annotation of peer schemas

in its local schema using terms from the domain ontology, as shown in Figure 4, where $Band\_P_1$ and $Band\_P_2$ two new classes defined as follows:

$$Band\_P_1 \ : \ Band \ \sqcap \ \geq_1 released$$
$$Band\_P_2 \ : \ Band \ \sqcap \ \forall type.Jazz \ \sqcap \ \geq_3 released \ \sqcap \ \forall year.(\leq 2000) \tag{2}$$

For simplicity, we have assumed a single relation for each peer in this example. The case of multiple relations linked with foreign keys is handled similarly: the class definition would contain an additional object property, corresponding to the foreign key, and having as range the class annotating the linked relation.

Exchanging data between $\mathcal{P}_1$ and $\mathcal{P}_2$ is meaningful and useful for both peers. However, information available at $\mathcal{P}_2$ is only partially sufficient for the information needs of $\mathcal{P}_1$, while, inversely, $\mathcal{P}_1$ contains much information that is irrelevant for the interests of $\mathcal{P}_2$. This obviously affects the quality of results that may be obtained by each peer. Notice also the *asymmetry* in the two directions. A qualitative measure is needed so that each peer may evaluate how suitable a particular acquaintance is, both in general, as well as with respect to a specific query.

3   Semantic Comparison of Peer Schemas

In this section, we propose a measure of the semantic similarity between the type of information provided by two peers, namely $Sem\_Sim(\mathcal{P}_1, \mathcal{P}_2)$. This measure is based on the use of the notions *recall* and *precision* adopted from the field of Information Retrieval. Using the pair of values $(recall, precision)$ to express the degree of semantic similarity between two peers, the measure proposed provides an intuitive way (a) to account for the asymmetry resulting from the specific direction considered in the comparison, and (b) to express the extend to which the information provided by a peer is a subset or superset of the requested information.

Recall and precision are widely used measures for evaluating the performance of Information Retrieval systems [12], which in general, are defined as follows.

- *Recall* is the proportion of relevant material actually retrieved in answer to a search request.
- *Precision* is the proportion of retrieved material that is actually relevant.

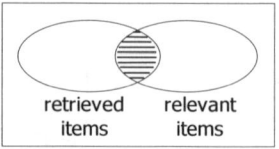

Fig. 5. Results for a search request

In Figure 5 the notion of both measures is pictorially depicted. For a given query, the values of recall and precision are calculated by

$$recall = \frac{|relevant \cap retrieved|}{|relevant|}$$
$$precision = \frac{|relevant \cap retrieved|}{|retrieved|} \tag{3}$$

A single measure combining recall and precision is the weighted harmonic mean, a.k.a. the F-measure. The general formula for non-negative real a is:

$$F_a = \frac{(1+a) * precision * recall}{a * precision + recall} \tag{4}$$

Choosing $a > 1$, it weights recall more than precision. In the literature, typical values for a are $0.5, 1, 2$. Although we may express the similarity measure in terms of the F-measure, we will refer to the measure proposed as a pair of values $(recall, precision)$; the two different metrics may be of different value and use to the peer. The latter can decide whether to employ the F-measure and determine appropriately the value of a.

The semantics of the peer's schema is made explicit by its annotation, which comprises a set of classes. Thus, instead of estimating the semantic similarity Sem_Sim $(\mathcal{P}_1, \mathcal{P}_2)$ between two peers $\mathcal{P}_1$ and $\mathcal{P}_2$, it suffices to estimate the similarity between the respective sets of classes in the ontology, i.e., $Sem\_Sim(\mathcal{C}_{\mathcal{P}_1}, \mathcal{C}_{\mathcal{P}_2})$. For simplicity, we first focus our study on the similarity between two classes and then, we extend the results for comparing two sets of classes. Hence, to adapt the notions of recall and precision in our context, we consider as relevant items the instances of the class in the semantic annotation of the requesting peer and as retrieved items the instances of the class in the semantic annotation of the provider peer. Therefore:

$$recall(C_{\mathcal{P}_1}, C_{\mathcal{P}_2}) = \frac{|\{x \mid x \in (C_{\mathcal{P}_1} \sqcap C_{\mathcal{P}_2})\}|}{|\{x \mid x \in C_{\mathcal{P}_1}\}|}$$
$$precision(C_{\mathcal{P}_1}, C_{\mathcal{P}_2}) = \frac{|\{x \mid x \in (C_{\mathcal{P}_1} \sqcap C_{\mathcal{P}_2})\}|}{|\{x \mid x \in C_{\mathcal{P}_2}\}|} \tag{5}$$

Notice that the above definitions for recall and precision have the following properties:

- When the two classes are equivalent, i.e., $C_{\mathcal{P}_1} \equiv C_{\mathcal{P}_2} \equiv C_{\mathcal{P}_1} \sqcap C_{\mathcal{P}_2}$, then $recall = 1$ and $precision = 1$, meaning that the contents of the two peers refer to the same type of entities.
- When $C_{\mathcal{P}_1} \sqsubseteq C_{\mathcal{P}_2}$, then $C_{\mathcal{P}_1} \sqcap C_{\mathcal{P}_2} \equiv C_{\mathcal{P}_1}$, thus $recall = 1$ and $precision < 1$, meaning that $\mathcal{P}_2$ contains information that is not of interest for $\mathcal{P}_1$.
- When $C_{\mathcal{P}_1} \sqsupseteq C_{\mathcal{P}_2}$, then $C_{\mathcal{P}_1} \sqcap C_{\mathcal{P}_2} \equiv C_{\mathcal{P}_2}$, thus $recall < 1$ and $precision = 1$, meaning that $\mathcal{P}_2$ can only partial cover the information needs of $\mathcal{P}_1$.
- When $\neg(C_{\mathcal{P}_1} \sqcap C_{\mathcal{P}_2} \sqsubseteq \bot)$, then $recall < 1$ and $precision < 1$, meaning that part of the information of each peer is of interest to the other.
- Finally, when $C_{\mathcal{P}_1} \sqcap C_{\mathcal{P}_2} \sqsubseteq \bot$, then $recall = 0$ and $precision = 0$. Essentially this means that the acquaintance with this peer is of no use.

Calculating the recall and precision from equations (5) requires knowledge about the extensions of the schemas, possibly using statistical techniques. However, in this paper, we focus on a different approach and calculate these measures based on the semantic information conveyed by the domain ontology, and in particular (a) the class hierarchy, (b) the property hierarchy, and (c) the restrictions on the properties of the classes. As the above approaches may be used together, our future plans include the improvement of the method by using both approaches in a combined way.

In the case of classes that are not described by any properties, the recall and precision can be measured by the ratio of their common ancestors. This is a common approach for measuring the similarity between classes in a taxonomy [13,14]. Thus, if $A(C)$ denotes the set of superclasses of a class C, then:

$$recall(C_1, C_2) = \frac{|A(C_1) \cap A(C_2)|}{|A(C_2)|}$$
$$precision(C_1, C_2) = \frac{|A(C_1) \cap A(C_2)|}{|A(C_1)|} \tag{6}$$

Notice that the values of recall and precision obtained from equations (6) adhere to the cases discussed previously.

Similarly, the recall and precision between two properties, p_1 and p_2, can be measured by the ratio of their common superproperties, as inferred by the property hierarchy in the domain ontology:

$$recall(p_1, p_2) = \frac{|A(p_1) \cap A(p_2)|}{|A(p_2)|}$$
$$precision(p_1, p_2) = \frac{|A(p_1) \cap A(p_2)|}{|A(p_1)|} \tag{7}$$

Concerning restrictions defined on the properties of a class, three different cases can be identified: (a) value restrictions on object properties (e.g., $\forall type.Jazz$); (b) value restrictions on datatype properties (e.g., $\forall year.(\leq 2000)$); and (c) cardinality restrictions (e.g., $\geq 3released$). In the first case, we consider the recall and precision between the two classes to which the values of the property are restricted. The other two cases can be handled by checking for a subsumption relationship between the two restrictions R_1

condition	$recall(R_1, R_2)$	$precision(R_1, R_2)$
$R_1 \equiv R_2$	1	1
$R_1 \sqsubseteq R_2$	1	0.5
$R_1 \sqsupseteq R_2$	0.5	1
$\neg(R_1 \sqcap R_2 \sqsubseteq \perp)$	0.5	0.5
$R_1 \sqcap R_2 \sqsubseteq \perp$	0	0

Fig. 6. Different cases of subsumption relationship between two restrictions R_1 and R_2

and R_2 as depicted in Figure 6. More accurate results may be obtained in the case that statistical knowledge about the value distributions of the underlying data is available.

Thus, equations (7) are updated to account also for the existence of restrictions, as follows:

$$recall(p_1, p_2) = \frac{|A(p_1) \cap A(p_2)|}{|A(p_2)|} \cdot \prod_{R(p_2)} recall(R'_i(p_1), R_i(p_2))$$

$$precision(p_1, p_2) = \frac{|A(p_1) \cap A(p_2)|}{|A(p_1)|} \cdot \prod_{R(p_1)} precision(R_i(p_1), R'_i(p_2)) \tag{8}$$

where $R(p)$ is the set of restrictions on property p, and $R(p)$, $R(p')$ denote a pair of corresponding restrictions; i.e., of the same type. The product is used in equations (8) as a decreasing monotonic function that captures the intuition that each factor contributes negatively to the final result. If no corresponding restriction is set on one of the compared properties, then, in the case of value restriction on object properties, the restricted range is compared to the default range of the property. In the other two cases, the value of recall or precision, accordingly, is set to a fixed value; the default is 0.5. Here, we do not consider the case of other intermediate values in the range $(0, 1)$ that may capture richer ranking semantics; e.g., $year < 2005$ may be preferable to $year < 2007$ for a request $year < 2004$; currently, both have precision equal to 0.5. (However, in that case one should deal with the issue of data range and distribution; a problem that we have left as a future extension to our approach.)

Therefore, given two classes C_1 and C_2, the recall and precision is calculated by adding the results for their individual properties and normalizing to the number of properties:

$$recall(C_1, C_2) = \frac{\sum_{P(C_2)} recall(p(C_1), p'(C_2))}{|P(C_2)|}$$

$$precision(C_1, C_2) = \frac{\sum_{P(C_1)} precision(p(C_1), p'(C_1))}{|P(C_1)|} \tag{9}$$

where $P(C)$ denotes the set of properties of class C, and $p(C_1)$, $p'(C_2)$ a pair of corresponding properties in the compared classes.

Example (Cont'd). We demonstrate the presented approach by applying the derived equations to estimate the semantic similarity between the schemas of the two peers, $\mathcal{P}_1$ and $\mathcal{P}_2$, of the motivating example introduced in Section 2.3. The schema of each peer comprises a single relation, semantically described by the definitions shown in the formulae (2). Given these definitions, and the ontology shown in Figure 3, from equations (9) follows that:

$$recall(\mathcal{P}_1, \mathcal{P}_2) = (3 \cdot 1 + 3 \cdot 0.5)/6 = 0.75$$
$$precision(\mathcal{P}_1, \mathcal{P}_2) = 1$$

The above results reflect, as expected, the fact that the type of information provided by peer $\mathcal{P}_2$ is more restricted w.r.t. that provided by peer $\mathcal{P}_1$.

The comparison can be extended to sets of classes, by comparing each class in the one set to its matching class in the other set (i.e., the class maximizing recall or precision, accordingly), and then normalizing to the cardinality of the sets. Therefore, for two sets of classes, $\mathcal{C}_1$ and $\mathcal{C}_2$, the following equations hold:

$$recall(\mathcal{C}_1, \mathcal{C}_2) = \frac{\displaystyle\sum_{C_i \in \mathcal{C}_1} \max_{C_j \in \mathcal{C}_2} recall(C_i, C_j)}{|\mathcal{C}_1|}$$

$$precision(\mathcal{C}_1, \mathcal{C}_2) = \frac{\displaystyle\sum_{C_j \in \mathcal{C}_2} \max_{C_i \in \mathcal{C}_1} precision(C_i, C_j)}{|\mathcal{C}_2|} \tag{10}$$

4   Extending to Queries and Mappings

So far, we have considered only the peers' semantic annotations, which refer to all the content potentially stored in the peer. Therefore, the previous analysis applies to the case of unrestricted exchange of information between two peers. However, the similarity needs to be evaluated also with respect to a specific query issued at a peer, as well as the mappings between two peers, which determine how the query is rewritten.

When a query Q_o is forwarded from $\mathcal{P}_1$ to $\mathcal{P}_2$, it is rewritten according to the mappings, resulting in a query Q_r. During the rewriting process, some attributes may not be rewritten, or may be approximately rewritten, while conditions may be lost or inserted, due to the nature of the specified mappings (e.g. due to value constraints in the mappings). As a result, the retrieved information may not completely adhere to the initial request. Therefore, the previous analysis needs to be extended to consider two additional factors:

- the portion of attributes that were rewritten and how accurate the rewriting was, and
- the conditions specified, both in the original query Q_o and its rewritten version Q_r.

Algorithm SSR

Input: The original query Q_o, issued at peer $\mathcal{P}_1 = (\mathcal{R}_1, \mathcal{O}, \mathcal{A}_1)$
     The target peer $\mathcal{P}_2 = (\mathcal{R}_2, \mathcal{O}, \mathcal{A}_2)$
     The mappings $M_{1,2}$ between the schemas of the two peers
Output: The recall and precision measures between the original query Q_o
     and the produced rewritten query Q_r

1. **Begin**
2. $\mathcal{R}_{Q_o} \leftarrow$ the set of relations appearing in Q_o
3. $\mathcal{C}_{Q_o} \leftarrow$ the set of classes annotating the relations in $\mathcal{R}_{Q_o}$
4. $\mathcal{C}_{Q_o,e} \leftarrow \mathcal{C}_{Q_o}$
5.   **Foreach** condition w in Q_o {
6. $R.t \leftarrow$ the attribute to which w is applied
7. $C \leftarrow$ the class corresponding to R
8. $P \leftarrow$ the property corresponding to t
9. $D \leftarrow$ the class or datarange representing the restricted value
10. $C \leftarrow C \sqcap \forall P.D$
11. $\mathcal{C}_{Q_o,e} \leftarrow$ update definition of C
12.   }
13. $Q_r \leftarrow rewrite(Q_o, M_{1,2})$
14. $\mathcal{C}_{Q_r,e} \leftarrow$ repeat lines 2-12 for Q_r
15. $recall = recall(\mathcal{C}_{Q_o,e}, \mathcal{C}_{Q_r,e})$
16. $precision = precision(\mathcal{C}_{Q_o,e}, \mathcal{C}_{Q_r,e})$
17. **End.**

Fig. 7. Algorithm for measuring the semantic similarity for rewritten queries

Any conditions existing in the query, either directly specified by the user or resulting from the mappings, further restrict the information requested or provided by the peer. Hence, these conditions need to be taken into account, together with the restrictions already existing in the classes semantically describing the peer's schema. For this purpose, each condition in the query is translated to a corresponding value restriction, which is added to the respective class in the peer's annotation. The process goes as follows:

1. A query Q_o is issued at peer $\mathcal{P}_1$.
2. Based on the peer's semantic annotations, the set of classes $\mathcal{C}_{Q_o}$ is selected, containing the classes annotating the relations in Q_o.
3. Each class in $\mathcal{C}_{Q_o}$ is enhanced with additional value restrictions on its properties, according to the conditions specified in Q_o, resulting in the set $\mathcal{C}_{Q_o,e}$.
4. Q_o is sent to peer $\mathcal{P}_2$, where it is rewritten as Q_r, according to the corresponding mappings.
5. Steps 2 and 3 are repeated for the rewritten version of the query and the semantic annotations of $\mathcal{P}_2$, resulting in the set of classes $\mathcal{C}_{Q_r,e}$.

Afterwards, the first step to evaluate the quality of the performed rewriting is to calculate the recall and precision between the two sets of classes $\mathcal{C}_{Q_o,e}$ and $\mathcal{C}_{Q_r,e}$. This is achieved by equations (10). The aforementioned process is formally described by the algorithm SSR, which is depicted in Figure 7.

The second factor to consider is the rewriting of the attributes appearing in the SE-LECT part of the query. If an attribute t was rewritten to an attribute t', then the quality of this rewriting is measured by the value of recall and precision between the corresponding properties p_t and $p_{t'}$ annotating these attributes. This is achieved by equations (7). The sum over all attributes is then calculated and normalized to the number of attributes in the query. If a SELECT attribute failed to be rewritten, then the value of recall for it is zero, as the corresponding information can not be retrieved. Precision is not affected, since no redundancy in the results is caused.

The results from the two factors are multiplied, since both contribute negatively to the quality of the performed rewriting:

$$recall(\mathcal{C}_{Q_o,e}, \mathcal{C}_{Q_r,e}, Q_o, Q_r) = \frac{\sum\limits_{t \in SELECT(Q_o)} recall(p_t, p_{t'})}{|SELECT(Q_o)|} \times recall(\mathcal{C}_{Q_o,e}, \mathcal{C}_{Q_r,e})$$

$$precision(\mathcal{C}_{Q_o,e}, \mathcal{C}_{Q_r,e}, Q_o, Q_r) = \frac{\sum\limits_{t' \in SELECT(Q_r)} precision(p_t, p_{t'})}{|SELECT(Q_r)|} \times precision(\mathcal{C}_{Q_o,e}, \mathcal{C}_{Q_r,e})$$

$$(11)$$

Example (Cont'd). We revisit the example of Section 2.3 to demonstrate two indicative cases.

Case 1. Assume that $\mathcal{P}_1$ is interested in bands formed at the 80's, so it issues the query:

$$Q_o \; : \; SELECT \; name, \; members, \; year \; FROM \; bands$$
$$WHERE \; year \geq 1980 \; AND \; year < 1990$$

Based on this query, the definition of $\mathcal{P}_1$ is updated as:

$$\mathcal{C}_{Q_o,e} \; : \; Band \; \sqcap \; \geq_1 released \; \sqcap \; \forall year.([1980, 1990))$$

The query is then forwarded to $\mathcal{P}_2$, and is rewritten as:

$$Q_r \; : \; SELECT \; name, \; singer, \; year \; FROM \; bands$$
$$WHERE \; year \geq 1980 \; AND \; year < 1990$$

Notice that the attribute $members$, corresponding to the property $hasMember$, has been rewritten as $singer$, corresponding to the property $hasSinger \sqsubseteq hasMember$. The definition of $\mathcal{P}_2$ is then updated accordingly:

$$\mathcal{C}_{Q_r,e} \; : \; Band \; \sqcap \; \forall type.Jazz \; \sqcap$$
$$\geq_3 released \; \sqcap \; \forall year.([1980, 1990))$$

Being more restrictive, the condition on attribute $year$ overwrites the previously existing restriction in the definition.

Applying equations (11) results in:

$$recall(\mathcal{P}_1, \mathcal{P}_2, Q_o, Q_r) = [(2 \cdot 1 + 0, 5)/3] \cdot [(4 \cdot 1 + 2 \cdot 0.5)/6] = 0.7$$
$$precision(\mathcal{P}_1, \mathcal{P}_2, Q_o, Q_r) = 1$$

Notice how the value of recall is affected negatively by the rewriting of $members$ to $singer$, and positively by the presence of the condition on attribute $year$ in the issued query, which counteracts the respective restriction in the annotation of $\mathcal{P}_2$.

Case 2. Assume now that the attribute $year$ was not present in $\mathcal{P}_2$'s schema or in the mapping. Then, the rewritten query may be:

$$Q_r \; : \; SELECT \; name, \; singer \; FROM \; bands$$

Since no conditions are present in the rewritten query, $\mathcal{P}_2$'s definition is not affected. Applying equations (11) results in:

$$recall(\mathcal{P}_1, \mathcal{P}_2, Q_o, Q_r) = [(1 + 0.5 + 0)/3] \cdot [(4 \cdot 1 + 2 \cdot 0.5)/6] = 0.42$$
$$precision(\mathcal{P}_1, \mathcal{P}_2, Q_o, Q_r) = 1 \cdot (5.5/6) = 0.92$$

Notice that the failure to rewrite the attribute $year$ significantly reduces the value of recall. Due to that failure, the returned bands have the restriction $year \leq 2000$, instead of the requested $year \in [1980, 1990)$, which negatively affects the value of precision.

5   Comparison of Peers and Queries with Multiple Ontologies

In the previous sections, we have presented a semantic similarity measure for peer schemas and queries propagated in the P2P network, assuming the existence of a single ontology that is used to annotate the elements of the schemas exposed by each peer. Even though this constitutes a valid assumption for several applications, maintaining an agreement to a common ontology, as the network size grows, becomes increasingly difficult. Therefore, it is important to consider how the proposed approach generalizes to the case of an environment where different peers may employ different ontologies to describe their schemas, as shown in Figure 2b. To maintain semantic interoperability in the absence of a common ontology, appropriate mappings need to be established between the terms used in the involved ontologies. In fact, a significant body of work exists addressing the issue of schema/ontology matching, as this constitutes a critical step in a variety of applications, such as data warehouses, catalog integration, agent communication, Web services coordination, and so on. In this section, we give an overview of the basic techniques underlying the state-of-the-art approaches for ontology matching and extend our similarity measure for comparing elements mapped from different ontologies.

Mapping Elements from Different Ontologies. A recent survey of ontology matching techniques is presented in [15]. As in [15], we consider as a mapping between the elements e and e' of two ontologies O and O', respectively, a tuple (id, e, e', M, c), where: id is a unique identifier for the particular mapping element; M is the relation between the elements e and e', i.e. equivalence ($\equiv$), subsumption ($\sqsubseteq$), overlap ($\sqcap$); and $c \in [0, 1]$ is a measure expressing the degree of confidence that M holds. Typically the matching process exploits, usually in a hybrid or composite way, techniques of the following types.

- *Element matching*. Similarity between elements is calculated based on their names and descriptions (i.e. labels, comments). Usually, a pre-processing occurs first, using Natural Language Processing techniques, such as tokenization (for example, $EmployeeRecord \rightarrow \langle Employ, Record \rangle$), lemmatization/stemming (e.g., $salaries \rightarrow salary$), and stopword elimination (e.g., removing articles or prepositions.) Then, string matching techniques are applied, such as prefix or suffix matching, edit distance, and n-grams (see [16] for a comparison of string matching techniques).
- *Structure matching*. In contrast to the previous case, where two elements are compared in isolation, these techniques consider the relations of these elements to other elements, using tree- or graph-matching techniques. For example, the similarity between two inner nodes can be calculated based on the similarity of their children. Alternatively, model-based approaches encode the intended semantics of each node, together with domain and structural knowledge, in a set of logical formulae, shifting the problem to one of logical satisfiability, that can be solved by employing standard SAT solvers.
- *Auxiliary information*. It is possible that the matcher uses external resources to obtain additional information to guide the matching process. Typical cases are the use of domain-specific thesauri, general-purpose dictionaries (e.g. WordNet [17]), repositories of previously mapped elements/structures, user feedback.

Translating Elements to a Different Ontology. Once appropriate mappings between two ontologies have been established, either manually, semi-automatically or automatically, these mappings can be used to merge the two ontologies or to translate elements from one ontology to the other. Examples of tools for ontology merging are OntoMerge [18] and PROMPT [19]. However, creating and maintaining a merged ontology incurs a significant overhead.

On the other hand, a translation service for OWL ontologies is presented in [20]. The translation relies on a provided mapping between the vocabularies of the two ontologies. Then, a class C_1 from the source ontology can be characterized as *strongly-translatable*, *equivalent*, *identical*, *weakly-translatable* or *approximately-translatable* to a class C_2 from the target ontology, depending on its name mapping and the *translatability* of its associated properties and restrictions.

For the translation process, we follow a similar approach, which however differs in two issues. First, instead of the above categorization of the translatability of a class, we need to derive a quantitative measure for the quality of the translation. Second, the aforementioned work translates classes from the source ontology to already existing classes from the second ontology. For example, assume that:

$$C_1 : Department \sqcap \geq_{20} hasEmployee \text{ and } C_2 : Team \sqcap \geq_{10} hasWorker$$

are two classes from the source and target ontology, respectively. Then, with respect to the name mappings:

$$(Department \rightarrow Team) \text{ and } (hasEmployee \rightarrow hasWorker)$$

C_1 is strongly-translatable to C_2. However, we translate a class expression by "rewriting" it according to the name mappings, i.e., replacing each term in the expression to its corresponding term in the target ontology, thus allowing the result of the translation to be a potentially new expression in the target ontology. Thus, in the previous example, the translated expression would be:

$$C'_1 : Team \sqcap \geq_{20} hasWorker.$$

The intuition behind that lies in the need to deal with the generated class expressions for annotating user queries.

Therefore, we consider a translation function f_T, that translates a (simple or complex) class or property from a source ontology O_1 to a target ontology O_2 as follows:

- a simple class or property is translated to the class or property specified by the corresponding mapping
- a complex class is translated by translating the terms in its definition; if a term is not translatable, the corresponding part of the expression is omitted.

As discussed at the beginning of this section, the considered mapping elements have the form (id, e, e', M, c), with $c \in [0, 1]$ and $M = (equivalent, more general, less general, overlap)$. (For more complex translations, in the presence of mappings which are themselves class expressions, a theoretical investigation is provided in [21].) Apparently, a translation may often result in loss of quality, either due to the (unintended) generalization or specialization of the original concept's meaning or due to the inability to translate (part of) it. Therefore, the quality of the translation has to be measured and taken into account. We use the measures of recall (r_T) and precision (p_T) for this purpose. Specifically, for a property p, given the corresponding mapping element (id, p, p', R, c), we measure the translation recall and precision by:

$$(r_T(p), \ p_T(p)) = ((0.5^x \cdot c, \ 0.5^y \cdot c)) \tag{12}$$

where

$$(x, \ y) = \begin{cases} (0, \ 0) & \text{if } M = \text{equivalent} \\ (1, \ 0) & \text{if } M = \text{more general} \\ (0, \ 1) & \text{if } M = \text{less general} \\ (1, \ 1) & \text{if } M = \text{overlap} \end{cases} \tag{13}$$

If an element is not translatable then both recall and precision are equal to zero. Notice that the value 0.5 used in equation (12) is a default value. Other values may be used, derived, for example, from knowledge of the application domain or the confidence level of the match provided by the matcher.

The measurement for a class is derived similarly, with the difference that the translatability of its properties, including potential restrictions, is also taken into account. Due to the way the translation is performed, cardinality restrictions, as well as value restrictions on datatype properties are always translatable, provided that the property on which the restriction is applied is translatable. For a value restriction R_p on an object property p (e.g. $\forall p.C$), the translatability of the restriction is dependent on the translatability of the class being the filler of the restriction (i.e. C), denoted by $\phi(R_p)$. Thus, the following equations hold.

$$
r_T(C) = 0.5^x \cdot c \cdot \frac{\displaystyle\sum_{P(C)} r_T(p) \cdot r_T(\phi(R_p))}{|P(C)|}
$$

$$
p_T(C) = 0.5^y \cdot c \cdot \frac{\displaystyle\sum_{P(C)} p_T(p) \cdot p_T(\phi(R_p))}{|P(C)|}
\tag{14}
$$

where (x,y) as in equation 13. Finally, when translating sets of classes, the quality of the translation can be assessed by the average quality of translation of the classes in the set:

$$
r_T(\mathcal{C}) = \frac{\displaystyle\sum_{C \in \mathcal{C}} r_T(C)}{|\mathcal{C}|} \quad , \quad p_T(\mathcal{C}) = \frac{\displaystyle\sum_{C \in \mathcal{C}} p_T(C)}{|\mathcal{C}|}
\tag{15}
$$

Comparing elements across ontologies. Given the above procedures for matching and translating elements between different ontologies, the next step is to extend the introduced similarity measure to cover these cases. In Section 3, we defined a measure of semantic similarity, in terms of the notions of recall and precision, for pairs of properties, classes, and sets of classes belonging in the same ontology (see equations (8), (9) and (10), respectively). In the following, we extend these functions (distinguished by the symbol $\neq$) so that they can be applied to elements from different ontologies. This is based on the observation that additional loss of quality may result due to the inaccurate translation (or no translation) of (parts of) the compared elements. Therefore, assuming that e_1 is an element (i.e., a property p, a class C or a set of classes $\mathcal{C}$) from the source ontology O_1, $e'_1 = f_T(e_1)$ its translation to the target ontology O_2, and e_2 the corresponding (i.e., the most similar) element of e'_1 in the target ontology, then:

$$
recall^{\neq}(e_1, e_2) = r_T(e_1) \cdot recall(e'_1, e_2)
$$

$$
precision^{\neq}(e_1, e_2) = p_T(e_1) \cdot precision(e'_1, e_2)
\tag{16}
$$

6 Related Work

The importance of semantics in P2P overlays has been apparent from the early stages of research in this field. One of the first works to consider semantics is [22], which suggests the construction of semantic overlay networks, i.e., SONs. Another work in the steps of [22] is [23], which suggests the dynamic construction of the interest-based shortcuts in order for peers to route queries to nodes that are more likely to answer them. Towards this end, the work in [24] and [25] exploits implicit approaches for discovering semantic proximity based on the history of query answering and the least recently used nodes. Additionally, SQPeer is an extensive work on PDMS that share RDF data and they localize the query patterns using views [26]. In the same spirit, our work focuses on overlays that share structured data and considers the problem of semantic similarity of schemas and queries.

Furthermore, semantics have been considered in the specialized field of structured P2P overlays. GridVine manages mapping of complex data and schemas of meta-data, specifically RDF [27]. The system allows schema inheritance and the creation and index of translation links that map pairs of schemas. Similarly, pSearch forms a structured semantic overlay [28]. Documents as well as queries are represented as semantic vectors. Both GridVine and pSearch base search efficiency on the structured form of the overlay, and, thus, their solution is not applicable to the semantic diversity problem in an unstructured P2P system. In contrast to these works, we consider the problem of semantic similarity in unstructured P2P overlays. Additionally, Bibster exploits ontologies in order to enable P2P sharing of bibliographic data [29]. Ontologies are used for importing data, formulating and routing queries and processing answers. Peers advertise their expertise and learn through ontologies about peers with similar data and interests. However, Bibster does not incorporate the ontology information into any kind of semantic similarity, as our work does.

Query similarity has been explored in several works in the recent past. Some of these works deal with keyword matching in the database environment [30,31] or with the processing of imprecise queries [32,33,34]. The work in [35] deals with attribute similarity, but focuses on numeric data and on conclusions about similarity that can be deduced from the workload. Furthermore, in [36] queries are classified according to their structural similarity; yet, the authors focus on features that differentiate queries with respect to optimization plans. The works in [37] and [4] deal with semantic similarity, which can be extracted from structural query features. Finally, the work in [38] considers the semantic similarity of schemas in a PDMS and proposes the creation of a distributed index that is used in order to route queries effectively. Nevertheless, these works do not consider the problem of both extracting and measuring semantic information of schemas. Our work fulfills this research gap and, moreover, considers the position and semantic evaluation of complex structured queries.

In [39] a notion of syntactic similarity is used to measure the extent to which a query is preserved after transformation. To achieve semantic interoperability in a bottom-up, semi-automatic manner, two feedback mechanisms are presented: one at the schema level, namely analyzing query translations along cycles in the network, and another at the data level, namely analyzing query results obtained through composite translations. This approach can be viewed as complementary to ours, as it can be used to incrementally develop global agreements among the participating peers.

In [40] a similarity measure for DL concepts is proposed. Concepts are represented in disjunctive normal form, and similarity is measured based on the overlap of these descriptions. The proposed similarity measure is symmetric for primitive concepts. For defined concepts it is asymmetric. However, similarity decreases only when the user's search concept is more specific than the examined one, while in the case that the examined concept is more specific than the requested by the user, similarity is not affected.

A graph-based semantic similarity is proposed in [41] to measure semantic relationships among pairs of Web pages, based on topical directories. The approach is based on the use of a weighting scheme to distinguish the role of different edges (e.g., hierarchical vs. non-hierarchical).

7 Conclusions

In this paper, we have dealt with that arise in P2P systems consisting of peer databases, i.e., peers that share structured data through the use of schema mappings. In such systems, information is requested by queries that are issued on local schemas and are rewritten to schemas of acquainted peers through mappings. Such a system can be enhanced with an ontology, which describes the domain of interest of the participating peers. We have discussed the semantic diversity between peer schemas, as well as between queries and their rewritten versions on other peers. We have chosen to rely on Semantic Web technology and, specifically, the use of domain ontologies, which enables peers to semantically annotate their elements despite the absence of a global schema. Using these ideas, we have proposed a similarity measure for schemas and queries based on the notions of recall and precision. The measure introduced takes into consideration the semantic annotations of schema elements and the structure and semantics of queries, as well as of the mappings used for the rewriting.

As future work, we intend to experiment on the proposed measure in order to identify semantically relevant peers, and evaluate the quality of the received answers to peer queries. A challenge is to adapt our method to a social network application and to evaluate its effectiveness in that domain. Social networks form naturally a P2P environment. We believe that the proposed approach would be beneficial in such a setting, as it would allow the participants to semantically express their interests, thus choosing their acquantancies based on their information needs. Furhtermore, we plan to investigate the notion of synopsis of similarity values, so that they can guide the propagation of queries to the most relevant peers in the overlay. Another direction to explore is the use of summaries and statistical techniques to extend the measure of similarity at the data level.

References

1. Arenas, M., Kantere, V., Kementsietsidis, A., Kiringa, I., Miller, R.J., Mylopoulos, J.: The hyperion project: from data integration to data coordination. SIGMOD Record 32(3), 53–58 (2003)
2. Halevy, A., Ives, Z., Suciu, D., Tatarinov, I.: Schema Mediation in Peer Data Management Systems. In: ICDE (2003)
3. Halevy, A.Y.: Answering Queries Using Views: A Survey. VLDB J 10(4), 270–294 (2001)
4. Kantere, et al.: V.: Coordinating P2P Databases Using ECA Rules. In: DBISP2P (2003)

5. Berners-Lee, T., Hendler, J., Lassila, O.: The Semantic Web. Scientific American 284(5), 34–43 (2001)
6. Borst, W.N.: Construction of Engineering Ontologies for Knowledge Sharing and Reuse. PhD thesis, University of Twente, Enschede, The Netherlands (1997)
7. McGuinness, D.L., van Harmelen, F.: OWL Web Ontology Language Overview. W3C Recommendation, W3C (February 2004),
   `http://www.w3.org/TR/2004/REC-owl-features-20040210/`
8. Baader, F., Calvanese, D., McGuinness, D.L., Nardi, D., Patel-Schneider, P.F. (eds.): The Description Logic Handbook: Theory, Implementation, and Applications. Cambridge University Press, Cambridge (2003)
9. Patel-Schneider, P.F., Horrocks, I.: OWL 1.1 Web Ontology Language. W3C Member Submission, W3C (December 2006)
10. Pan, J.Z., Horrocks, I.: OWL-Eu: Adding customised datatypes into owl. In: Gómez-Pérez, A., Euzenat, J. (eds.) ESWC 2005. LNCS, vol. 3532. Springer, Heidelberg (2005)
11. Lenzerini, M.: Data Integration: A Theoretical Perspective. In: PODS, pp. 233–246 (2002)
12. Baeza-Yates, R.A., Ribeiro-Neto, B.A.: Modern Information Retrieval. ACM Press / Addison-Wesley (1999)
13. Lin, D.: An Information-Theoretic Definition of Similarity. In: ICML (1998)
14. Resnik, P.: Using Information Content to Evaluate Semantic Similarity in a Taxonomy. In: IJCAI, pp. 448–453 (1995)
15. Shvaiko, P., Euzenat, J.: A survey of schema-based matching approaches. J. Data Semantics IV, 146–171 (2005)
16. Cohen, W.W., Ravikumar, P., Fienberg, S.E.: A comparison of string metrics for matching names and records. In: Proceedings of the KDD-2003 Workshop on Data, Washington, DC, pp. 13–18 (2003)
17. Miller, G.A.: Wordnet: a lexical database for english. Commun. ACM 38(11), 39–41 (1995)
18. Dou, D., McDermott, D.V., Qi, P.: Ontology translation on the semantic web. J. Data Semantics 2, 35–57 (2005)
19. Noy, N.F., Musen, M.A.: Prompt: Algorithm and tool for automated ontology merging and alignment. In: AAAI/IAAI, pp. 450–455. AAAI Press / The MIT Press (2000)
20. Mota, L., Botelho, L.: Owl ontology translation for the semantic web. In: Proceedings of the Semantic Computing Workshop of the 14th International World Wide Web Conference (2005)
21. Beeri, C., Levy, A.Y., Rousset, M.C.: Rewriting queries using views in description logics. In: PODS, pp. 99–108. ACM Press, New York (1997)
22. Crespo, A., Garcia-Molina, H.: Semantic Overlay Networks for P2P Systems. In: Moro, G., Bergamaschi, S., Aberer, K. (eds.) AP2PC 2004. LNCS (LNAI), vol. 3601, pp. 1–13. Springer, Heidelberg (2005)
23. Sripanidkulchai, K., Maggs, B., Zhang, H.: Efficient Content Location Using Interest-Based Locality in Peer-to-Peer Systems. In: INFOCOM (2003)
24. Voulgaris, S., et al.: Exploiting Semantic Proximity in Peer-to-Peer Content Searching. In: FTDCS (2004)
25. Handurukande, S., et al.: Exploiting Semantic Clustering in the eDonkey P2P Network. In: ACM SIGOPS (2004)
26. Kokkinidis, G., Sidirourgos, E., Christophides, V.: Query Processing in RDF/S-based P2P Database Systems. In: Semantic Web and Peer-to-Peer, pp. 59–81. Springer, Heidelberg (2006)
27. Aberer, K., Cudre-Mauroux, P., Hauswirth, M., Pelt, T.V.: Gridvine:Building internet-scale semantic overlay networks. In: McIlraith, S.A., Plexousakis, D., van Harmelen, F. (eds.) ISWC 2004. LNCS, vol. 3298, pp. 107–121. Springer, Heidelberg (2004)

28. Tang, C., et al.: Peer-to-Peer Information Retrieval Using Self-Organizing Semantic Overlay Networks. In: SIGCOMM (2003)
29. Haase, P., et al.: Bibster - A Semantics-based Bibliographic Peer-to-Peer System. Journal of Web Semantics 2(1), 99–103 (2005)
30. Agrawal, S., Chaudhuri, S., Das, G.: DBXplorer: A System for Keyword-Based Search over Relational Databases. In: ICDE (2002)
31. Cohen, W.: Integration of Heterogeneous Databases Without Common Domains Using Queries Based on Textual Similarity. In: SIGMOD (1998)
32. Fuhr, N.: A Probabilistic Framework for Vague Queries and Imprecise Information in Databases. In: VLDB (1990)
33. Kießling, W., Kostner, G.: Preference SQL - Design, Implementation, Experiences. In: VLDB (2002)
34. Motro, A.: VAGUE: A User Interface to Relational Databases that Permis Vague Queries. In: TOIS, vol. 6(3), pp. 187–214 (1988)
35. Agrawal, S., Chaudhuri, S., Das, G., Gionis, A.: Automated Ranking of Database Query Results. In: CIDR (2003)
36. Ghosh, A., Parikh, J., Sengar, V.S., Haritsa, J.R.: Plan Selection based on Query Clustering. In: VLDB (2002)
37. Chu, W.W., Zhang, G.: Associative Query Answering via Query Feature Similarity. In: IIS (1997)
38. Mandreoni, F., Martoglia, R., Sassateli, S., Penzo, W.: SRI: Exploitng Semantic Information for Effective Query Routing in a PDMS. In: WIDM (2006)
39. Aberer, K., Cudré-Mauroux, P., Hauswirth, M.: Start making sense: The chatty web approach for global semantic agreements. J. Web Sem. 1(1), 89–114 (2003)
40. Janowicz, K.: Sim-dl: Towards a semantic similarity measurement theory for the description logic cnr in geographic information retrieval. In: Meersman, R., Tari, Z., Herrero, P. (eds.) OTM 2006 Workshops. LNCS, vol. 4278, pp. 1681–1692. Springer, Heidelberg (2006)
41. Maguitman, A.G., et al.: Algorithmic computation and approximation of semantic similarity. In: World Wide Web, vol. 9(4), pp. 431–456 (2006)

Maintaining Semantic Mappings between Database Schemas and Ontologies

Yuan An[1] and Thodoros Topaloglou[2]

[1] College of Information Science and Technology
Drexel University, USA
yan@ischool.drexel.edu
[2] Department of Mechanical and Industrial Engineering
University of Toronto, Canada
thodoros@mie.utoronto.ca

Abstract. There is a growing need to define a semantic mapping from a database schema to an ontology. Such a mapping is an integral part of the data integration systems that use an ontology as a unified global view. However, both ontologies and database schemas evolve over time in order to accommodate updated information needs. Once the ontology and the database schema associated with a semantic mapping evolved, it is necessary and important to maintain the validity of the semantic mapping to reflect the new semantics in the ontology and the schema. In this paper, we propose a formulation of the mapping maintenance problem and outline a possible solution using illustrative examples. The main points of this paper are: (1) to differentiate the semantic mapping maintenance problem from the schema mapping adaptation problem which only adapts mappings when schemas change; (2) to develop an approach for specifying the validity of a semantic mapping in terms of two-way legal instances translation between two models; (3) to explore the approach of using simple correspondences to capture changes to ontologies/schemas; and (4) to sketch a solution using examples.

1 Introduction

A semantic mapping from a database schema to an ontology defines a semantic relationship between the schema and the ontology. For example, a many-to-many relationship between a concept C_1 and a concept C_2 in an ontology may be mapped to relational tables storing attributes of C_1 and C_2 and a linking table that maintains the association of the identifiers[1] of C_1 and C_2. Such a semantic mapping can be expressed in a declarative language that encodes the formal semantics of the schemas. In recent years, we are witnessing a growing demand for defining semantic mappings from database schemas to ontologies. For example, semantic mappings are integral part of ontology-based information integration systems [8,13], and data integration efforts in the context of the

---

[1] We assume that a subset of attributes of a concept in an ontology acts as identifier of the concept.

V. Christophides et al. (Eds.): SWDB-ODBIS 2007, LNCS 5005, pp. 138–152, 2008.

Semantic Web. Furthermore, a recent work [1] suggests that the semantics of database schemas expressed in terms of semantic mappings from schemas to conceptual models/ontologies provide opportunities to improve the capabilities of traditional schema mapping tools, even when different database schemas are associated with different conceptual models or ontologies.

However, both ontologies and schemas change over time in order to accommodate new information needs. Such change may cause an existing semantic mapping *invalid*. Therefore, once a semantic mapping from a schema to an ontology has been created, it is important and necessary to automatically, at least to some extent, maintain the validity of the semantic relationship when the schema and ontology evolve. We call this process *maintaining semantic mappings under evolution* or *mapping maintenance* for short. A typical solution to the mapping maintenance problem is to regenerate the semantic mapping between the evolved ontology and schema. The problem of the mapping regeneration solution is that the solution can be costly in terms of human effort and expertise. The reason is that semantic mapping creation is a demanding task which requires huge amount of human effort, because both the schema and the ontology that are related by a semantic mapping are complex artifacts which may contain hundreds and thousands modeling constructs. There are existing methods and tools, e.g., [3,2], for creating semantic mappings from database schemas to ontologies. But almost all the current tools are semi-automatic and interactive, requiring humans involved in the process. A better solution to the mapping maintenance problem is to incrementally update the existing semantic mapping to reflect *changes* in the ontology or schema. In this paper, we report on our preliminary study on the problem of incrementally maintaining a special type of semantic mapping, which, in a local-as-view fashion, relates a single atom (e.g., a table) in a schema with a conjunctive formula encoding a substructure in an ontology. The formalism is presented in Section 3.

The aims of the maintenance are two-fold: first, to preserve the semantic relationship between the schema and the ontology when the schema and ontology are modified; second, to reuse the existing semantic mapping as much as possible. A similar problem has been studied for adapting schema mappings under schema evolution. Two possible approaches are proposed in the literature: a schema change approach (SCA) [15] and a mapping composition approach (MCA) [16]. Both solutions focus on reusing the semantics encoded in previous mappings for merely adapting the mappings. Schemas are not updated accordingly. In our situation, adapting the ontology/schema associated with a semantic mapping along with the mapping will be essential for achieving desired goals. Consider a very simple case. Suppose the semantics of a relational database schema is expressed in terms of an ontology. If the database engineer wants to modify the schema by adding a new column to a table representing a concept in the ontology, it may be desirable to add a new attribute to the concept in the ontology in order to maintain the semantic relationship that covers the new element of the schema. Maximizing the coverage over schema will be one of the desired goals for maintaining a semantic mapping.

Although mapping maintenance is important and necessary for many applications, solutions to the problem are rare. This is due to many challenges involved, including: how to define validity/consistency of mapping and detect inconsistency of a mapping; what is a right mapping language; how to capture changes to ontologies and database schemas; how to devise a plan for updating mappings according to the intent and expectation of the user; and what are the principles for a systematic maintenance solution.

In this paper, we formulate the maintenance problem. We propose a specification for the validity of a semantic mapping. Subsequently, we describe the desired goals for maintaining semantic mappings between database schemas and ontologies, and we outline our solution for addressing the problem using a comprehensive set of examples.

The rest of the paper is organized as follows. Section 2 discusses related work. Section 3 introduces our formalism for a semantic mapping from a schema to an ontology. Section 4 characterizes schema and ontology evolution. Section 5 outlines a solution to the problem of semantic mapping maintenance. Finally, Section 6 concludes this paper.

2   Related Work

The directly related work is the study on schema mapping adaptation [15,16]. The goal of schema mapping adaptation is to automatically update a schema mapping by reusing the semantics of the original mapping when the associated schemas change. Yu & Popa [16] explore the schema mapping composition approach. Schema evolutions are captured by formal and accurate schema mappings, and schema adaptation is achieved by composing the evolution mapping with the original mapping. On the other hand, the schema change approach in [15] proposed by Velegrakis et al. incrementally changes mappings each time a primitive change occurs in the source or target schemas. Both solutions focus on reusing the semantics encoded in existing mappings for merely adapting the mappings without considering the synchronization between schemas. This is due to the nature of their problems where schema mappings are primarily used for *data exchange* [10], i.e., translating a data instance under a source schema to a data instance under a target schema. If a schema mapping connecting two schemas which are semantically inconsistent, then the data exchange process simply does not always produce a target instance. Our approach is different from these solutions in that we aim to maintain the *semantic validity* of semantic mappings through incremental updates on the mappings as well as associated ontologies/schemas.

Other related work includes schema evolution in object-oriented databases (OODB). The problem of schema evolution in OODB is to maintain the consistency of an OODB when its schema is modified. The challenges are to update the database efficiently and minimize information loss. A variety of solutions, e.g., [6,5,9,12], have been proposed in the literature. Our problem is different from the schema evolution problem in OODB in that we aim at the semantic consistency between a schema and an ontology. However, we can draw some insights from the extensive study of the schema evolution problem in OODB. In AutoMed [7,11], schema evolution and integration are combined in one unified framework. Source schemas are integrated into a global schema by applying a sequence of primitive transformations to them. The same set of primitive transformations can be used to specify the evolution of a source schema into a new schema. In our approach, we do not ask users to specify a sequence of transformations.

Another mapping maintenance problem studied in [14] mainly focus on detecting inconsistency of simple correspondences between schema elements when schemas evolve. This problem is complementary to the problem we consider here.

3   Semantic Mappings between Ontologies and Schemas

3.1   Relational Schemas and Ontologies

Here we focus on relational schemas described in the relational model. The basic data representation construct of the relational model is relation, which consists of a set of tuples. The schema of a relation or a *table* specifies the name of the relation, the name of each column (or attribute or field), and the type of each column. Furthermore, we can make the description of the collection of data more precise by specifying *integrity constraints*, which are conditions that the tuples in a table must satisfy. Here, we consider *key* and *foreign key* (abbreviated as *f.k.* henceforth) constraints. A key in a table is a subset of the columns of the table that uniquely identifies a tuple. A f.k. in a table T is a set of columns F that *references* the key of another table T' and imposes a constraint that the projection of T on F is a subset of the projection of T' on the key of T'. A relational schema thus consists of a set of relational tables and a set of key and f.k. constraints. Formally, we use the notation $T(k_1, k_2, ..., k_n, y_1, y_2, ..., y_m)$ to represent a relational table T with key $K = (k_1, k_2, ..., k_n)$.

An ontology describes a subject matter in terms of concepts, relationships, and attributes. In this study, we do not restrict ourselves to any particular language for describing ontologies. Instead, we use a generic conceptual modeling language (CML) which has the following features. The language allows the representation of *classes/concepts/entities* (unary predicates over individuals), *object properties/relationships* (binary predicates relating individuals), and *datatype properties/attributes* (binary predicates relating individuals with values such as integers and strings); attributes are single valued in this paper. Concepts are organized in the familiar ISA hierarchy. Relationships and their inverses (which are always present) are subject to constraints such as specification of domain and range, plus cardinality constraints of the form $k..l$; if the lower bound, $k = 1$, the relationship is called , *total*, if the upper bound, $l = 1$, the relationship is called *functional*. In addition, a subset of attributes of a concept is specified as the identifier of the concept. As in the Entity-Relationship model, a strong entity has a global identifier, while a weak entity is identified by an identifying relationship plus a local identifier. An ontology thus contains a set of concepts, relationships, and attributes as well as a set of identification and cardinality constraints.

We can represent a given ontology using a labeled directed graph, called an *ontology graph*. We construct the ontology graph from an ontology by considering concepts as nodes and relationships as edges. A many-to-many relationship p between concepts C and D will be written in text as $\boxed{C}$ ---p--- $\boxed{D}$. It will be important for our approach to distinguish *functional edges* – ones with upper bound cardinality of 1, and their composition: *functional paths*. If the relationship p is functional from C to D, we write $\boxed{C}$ ---p->-- $\boxed{D}$.

3.2   Semantic Mappings between Ontologies and Schemas

In this study, we use the semantic mapping notion that is proposed in [4] which relates tables in a schema with formulas over an ontology. The formula over an ontology is in a subset of conjunctive formulas and encodes a subtree in the ontology graph. In particular, we assume that the semantics of a table is represented by a subtree (subgraphs can be transformed into subtrees by duplicating nodes in cycles). We call such a subtree a *semantic tree (or s-tree)*, where columns of the table associate uniquely with attribute of the concepts in the s-tree. This assumption also corresponds to the standard database design practice where each table is derived from a structure, usually, a subtree, in a conceptual model. After encoding s-trees in conjunctive formulas by using unary predicates for concepts, binary predicates for attributes, and binary predicates for binary relationships (see [4]), we can represent a semantic mapping between a relational schema and an ontology using a set of formula of the form $T(X) \leftrightarrow \Phi(X, Y)$, where T is a table with columns X and Φ is a conjunctive formula over predicates representing an s-tree. X and Y are quantified variables as specified later.

Example 1. Gene expression databases maintain information on genes, biological samples and measurements on genes over samples. Biological sample is a central concept being modeled in a gene expression database. To record information about a sample which can be a tissue, cell, or RNA material that originates from a donor of a given species, one needs to create a sub-schema that we will refer to as the sample database (SDB). Suppose that a SDB contains a table

sample(sample_ID, species, organ, pathology,..., donor_ID),

where the underlined column sample_ID is the key of the table and donor_ID is a foreign key to a table called donor.

The semantics of the sample table can be expressed in terms of an s-tree in an ontology as shown in Figure 1 which is described in the UML notation, where identifier of a concept is indicated by the keyword key. The s-tree contains two concepts, SAMPLE and PERSON, and a relationship, originates, between the two concepts.

Graphically, we use dashed double-arrows to indicate the correspondences between columns of the relational table and attributes of concepts in the ontology. The correspondences plus the s-tree gives rise to a semantics of the table. Furthermore, the semantics of the table is expressed in the following formula

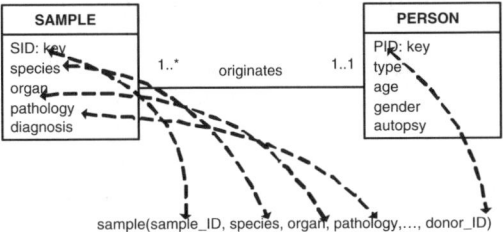

Fig. 1. The sample table and Its Semantics

sample($sample\_ID$, $species$,..., $donor\_ID$) $\leftrightarrow$
    SAMPLE(x_1), SID(x_1, $sample\_ID$), species(x_1, $species$),
    ..., PERSON(x_2), originates(x_1, x_2), PID(x_2, $donor\_ID$).   ■

Valid Semantic Mappings. Given a semantic mapping formula $T(X) \leftrightarrow \Phi(X, Y)$ which relates a table $T(X)$ in a schema with a conjunctive formula $\Phi(X, Y)$ encoding an s-tree $\mathcal{G}$ in an ontology. We say that $T(X) \leftrightarrow \Phi(X, Y)$ is valid if the table and the s-tree $\mathcal{G}$ are "semantically compatible". More specifically, we define the validity by using two logical formulas $\forall X(T(X) \rightarrow \exists Y.\Phi(X, Y))$ and $\forall X, Y(\Phi(X, Y) \rightarrow T(X))$, plus the key and f.k. constraints of the schema and the identification and cardinality constraints of the ontology.

The formula $\forall X(T(X) \rightarrow \exists Y.\Phi(X, Y))$ can be considered as the formal specification for translating instances from the table to the s-tree, and the formula $\forall X, Y(\Phi(X, Y) \rightarrow T(X))$ can be considered as the formal specification for translating instances from the s-tree to the table. Let Σ_T be the set of key and f.k. constraints of T. Let Σ_S be the set of identification and cardinality constraints of S. An instance I of T is a legal instance if I satisfies all constraints in Σ_T. An instance J of S is a legal instance if J satisfies all constraints in Σ_S. For each legal instance I of T, we can generate an instance J' of S through $\forall X(T(X) \rightarrow \exists Y.\Phi(X, Y))$ by instantiating Y. For each legal instance J of S, we can generate an instance I' of T through $\forall X, Y(\Phi(X, Y) \rightarrow T(X))$. If both J' and I' are legal instances of S and T, respectively, then we say that T and S are "semantically compatible."

We now define a valid semantic mapping using semantically compatible instances. Specifically, a formula $T(X) \leftrightarrow \Phi(X, Y)$ relating a table $T(X)$ in a schema with a conjunctive formula $\Phi(X, Y)$ encoding an s-tree $\mathcal{G}$ in an ontology is a **valid** semantic mapping formula, if and only if for each legal instance of T, we can generate a *legal* instance of $\mathcal{G}$ through $\forall X(T(X) \rightarrow \exists Y.\Phi(X, Y))$, and for each legal instance of $\mathcal{G}$, we can generate a *legal* instance of T through $\forall X.Y(\Phi(X, Y) \rightarrow T(X))$.

Having the definition about a valid semantic mapping formula, we attempt to (semi-)automatically maintain the validity of each formula when the schema and the ontology related by the formula evolve. In the next section, we begin with a characterization of possible changes in schemas and ontologies.

4   Evolution of Schemas and Ontologies

Changes to schemas and ontologies can be characterized by mappings [16] or by sequences of evolution primitives [15,5]. Consider a mapping $\mathcal{M}$ between two schema $\mathcal{S}_1$ and $\mathcal{S}_2$. If one of the schemas, e.g., $\mathcal{S}_1$, evolves to a new schema $\mathcal{S}'_1$, the mapping composition approach (MCA) for schema mapping adaptation will compose the mapping $\mathcal{M}$ with an evolution mapping $\mathcal{M}'$ between $\mathcal{S}'_1$ and $\mathcal{S}_1$ to derive a new mapping between $\mathcal{S}'_1$ and $\mathcal{S}_2$, while the schema change approach (SCA) will look at a sequence of primitive changes for adapting $\mathcal{M}$.

Both MCA and SCA approaches are inadequate in dealing with the problem of maintaining a semantic mapping $\mathcal{M}$ between a schema $\mathcal{S}$ and an ontology $\mathcal{O}$. First of all, neither MCA nor SCA approach attempts to maintain the validity of the semantic mapping. For example, if the key information of a table in $\mathcal{S}$ changes, the mapping $\mathcal{M}$ may not change, but the ontology $\mathcal{O}$ may need to be modified in order to keep $\mathcal{M}$ as a valid semantic mapping. However, current MCA and SCA approaches only consider mapping adaptation. Second, the MCA does not capture the changes of adding elements to schemas. If an element is added, it will leave the existing mapping unchanged. Third, it is not guaranteed that the current SCA approach would maintain the semantics of the existing mapping by using a sequence of primitive changes, as the set of primitive changes for schema evolution may not cover some changes encountered in ontology evolution. For example, one of primitive changes that may happen in ontology evolution but are not captured by the set of primitive changes for schema evolution is adding/deleting an ISA relationship between two concepts.

In this study, we use a set of correspondences to link elements of the previous schema/ontology to elements of the new schema/ontology when a schema/ontology changes. We then analyze the existing semantic mapping and the semantics in the new schema/ontology. Through the set of correspondences, we will then (semi-)automatically adapt both the semantic mapping and the schema/ontology to maintain the validity of the semantic mapping.

Example 2. Figure 2 depicts on the left an old ontology $\mathcal{O}_1$ consisting of a single concept BIOSAMPLE. On the right is the new ontology $\mathcal{O}'_1$ which was evolved from $\mathcal{O}_1$ by adding a new concept TISSUE. The dashed double-arrows from attributes of the BIOSAMPLE concept in $\mathcal{O}_1$ to attributes of the BIOSAMPLE and TISSUE concepts in $\mathcal{O}'_1$ capture the relationship between the old ontology and the new ontology.   ∎

Changes to schemas and ontologies can be classified along two orthogonal axes. First, on the *action* axis, changes can be classified into (1) changes for adding/deleting elements; (2) changes for merging/splitting elements; (3) changes for moving/copying elements; (4) changes for renaming elements; and (5) changes for modifying constraints. Second, on the *effect* axis, changes can be classified into (i) changes that cause mapping modification; (ii) changes that cause the related schema (or ontology) modification; and (iii) changes that cause both mapping and the related schema (or ontology) modification. The classification along the *effect* axis is mainly concerned with maintaining the validity of a

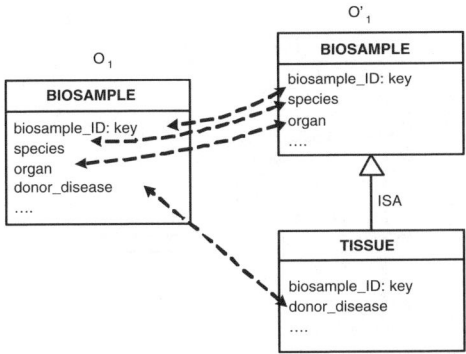

Fig. 2. The Correspondences between Old and New Ontologies

semantic mapping as specified in Section 3. In the next section, we discuss our solution to the maintenance problem by associating changes classified along the *action* axis with changes classified along the *effect* axis.

5   Maintaining the Semantic Mappings

We outline an algorithm for maintaining semantic mappings between relational schemas and ontologies. The input to the algorithm consists of a relational schema $\mathcal{S}$, an ontology $\mathcal{O}$, an existing valid semantic mapping $\mathcal{M}$ between $\mathcal{S}$ and $\mathcal{O}$, a new schema $\mathcal{S}'$ (or ontology $\mathcal{O}'$) evolved from $\mathcal{S}$ (or $\mathcal{O}$), and a set of element correspondences $\mathcal{M}'$ between $\mathcal{S}$ and $\mathcal{S}'$ (or between $\mathcal{O}$ and $\mathcal{O}'$). The output of the algorithm is an ontology $\mathcal{O}''$ (or a schema $\mathcal{S}''$) and the semantic mapping $\mathcal{M}''$ between $\mathcal{S}'$ and $\mathcal{O}''$ (or between $\mathcal{O}'$ and $\mathcal{S}''$). The ontology $\mathcal{O}''$ may be just the original ontology $\mathcal{O}$, if the schema $\mathcal{S}$ evolved and only the semantic mapping gets adapted without any changes in the original ontology. Similarly, the schema $\mathcal{S}''$ may be just the original schema $\mathcal{S}$, if what evolved was the ontology $\mathcal{O}$ and there are no needs to change the original schema in order to adapt the semantic mapping.

Figure 3 graphically describes the semantic mapping maintenance settings. Figure 3 (a) shows the situation where the schema $\mathcal{S}$ evolved to a new schema $\mathcal{S}'$. $\mathcal{M}$ is the existing semantic mapping; $\mathcal{M}'$ is the set of correspondences from elements of $\mathcal{S}'$ to elements of $\mathcal{S}$. The aim of the mapping maintenance is to adapt $\mathcal{M}$ to a new semantic mapping $\mathcal{M}''$ between $\mathcal{S}'$ and $\mathcal{O}$ (or $\mathcal{O}''$ if the original ontology needs to be modified.) Likewise, Figure 3 (b) show the situation when the ontology $\mathcal{O}$ evolved to $\mathcal{O}'$, $\mathcal{M}$ needs to be adapted to $\mathcal{M}''$ between $\mathcal{S}$ (or $\mathcal{S}''$) and $\mathcal{O}'$.

The maintenance algorithm is based on the knowledge about the existing semantic mapping and the analysis of the semantics in the changes to the schema/ontology. We first explore the knowledge encoded in a semantic mapping as studied in the previous work for discovering semantic mappings from schemas

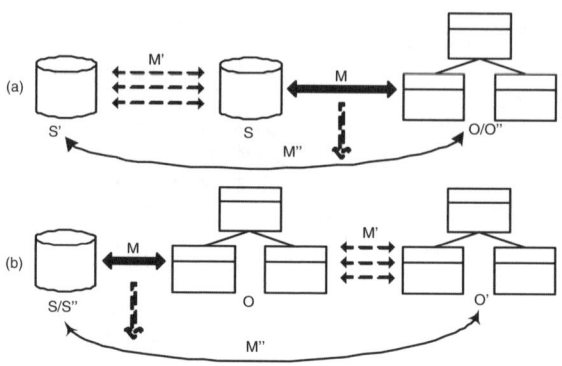

Fig. 3. Maintenance of Semantic Mapping

to ontologies [4]. Then we illustrate the algorithm by analyzing the semantics in changes using means of examples.

A semantic mapping formula $T(X) \leftrightarrow \Phi(X, Y)$ associates a table $T(X)$ with an s-tree in an ontology. There is additional knowledge about the association/ relationship [4]. Specifically, an s-tree can be decomposed into several skeleton trees: a skeleton tree corresponding to the key of the table, skeleton trees corresponding to f.k.s of the table, and skeleton trees corresponding to the rest of the columns of the table. Each skeleton tree has an anchor concept which is the root of the skeleton tree. To satisfy the semantics of the key in a table, the s-tree is connected by functional paths from the anchor of the key skeleton tree to the anchors of f.k. skeleton trees and other skeleton trees.

Example 3. Figure 4 shows a table sample(sid,tid,donor) storing the information about a sample, where sid is the sample identifier, tid is the identifier of the test that screens the sample, and donor is the identifier of the person donating the sample. The concept SAMPLE is modeled as a weak entity owned by the TEST concept. Therefore, the key of the sample table is the combination of the key of a table for the TEST concept and the local identifier sid. In addition, the donor column is a f.k. referencing the key of a table for the PERSON concept.

The semantics of the sample table is represented in terms of the s-tree above it in Figure 4. This s-tree consists of the skeleton tree $\boxed{\text{SAMPLE}}$ `---screenedIn->--` $\boxed{\text{TEST}}$ for the key of the sample table and the skeleton tree $\boxed{\text{PERSON}}$ for the f.k. of the sample table. The anchor of the key skeleton tree is the concept SAMPLE,

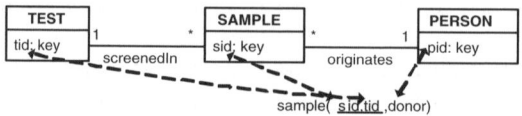

Fig. 4. Skeleton Trees in a Semantic Mapping

while the anchor of the f.k. skeleton tree is the concept **PERSON**. The s-tree is connected by a functional edge **originates** from the anchor **SAMPLE** to the anchor **PERSON**.   ∎

Each s-tree in a semantic mapping that consists of a key skeleton tree corresponding to the **key** of the table, skeleton trees corresponding to f.k.s of the table, skeleton trees corresponding to other columns of the table, and functional paths from the anchor of the key skeleton tree to anchors of other skeleton trees. To maintain a semantic mapping when the schema/ontology changes, we aim at maintaining the s-tree associated with the table. Our goals for maintaining semantic mappings are as follows.

Goal 1. *For a valid semantic mapping $\mathcal{M}$ between a schema $\mathcal{S}$ and an ontology $\mathcal{O}$, if $\mathcal{M}$ has been adapted to $\mathcal{M}''$ after some changes to the schema/ontology, then each mapping formula $m \in \mathcal{M}''$ must be a valid semantic mapping formula as specified in Section 3.*

Goal 2. *For a valid semantic mapping $\mathcal{M}$ between a schema $\mathcal{S}$ and an ontology $\mathcal{O}$, if $\mathcal{M}$ has been adapted to $\mathcal{M}''$ after some changes to the schema/ontology, then for each element $e \in \mathcal{S}$ that was covered by $\mathcal{M}$ (i.e., e was referred to by some mapping formulas in $\mathcal{M}$), e is covered by $\mathcal{M}''$ if e was not deleted from $\mathcal{S}$, and for each new element e' added to $\mathcal{S}$, e' is also covered by $\mathcal{M}''$.*

The first goal specifies the fundamental requirement for semantic mapping maintenance, that is, to maintain the validity of a semantic mapping according to the definition. The second goal requires that the semantic mapping after adapted should cover as much the remaining schema as covered by the existing semantic mapping and cover any newly added elements. The second goal comes from our intention of using semantic mappings for expressing semantics for database schemas. That is, for a database schema, we do not want to lose semantic information expressed in terms of the semantic mapping from the schema to an ontology after the semantic mapping is adapted due to changes to the schema/ontology.

The following examples outline the mapping maintenance algorithm in an intuitive way. The complete algorithm will be available in a full paper. At the present, the maintenance algorithm focuses on a pair of a schema and an ontology that are related by a semantic mapping.

Example 4. The following semantic mapping formula relates a relational table sample(<u>sid</u>,donor) with an s-tree in an ontology as shown in Figure 5 (a):

> sample(sid, $donor$) ↔
>> SAMPLE(x_1), sid(x_1, sid), PERSON(x_2),
>> originates(x_1, x_2), pid(x_2, $donor$).

First, we consider changes that add new column(s) to the relational table.

(1) Add a column that is neither part of the key nor a f.k.. For example, a new column **species** was added to the **sample** table. In this case, the algorithm will suggest to add a new attribute to the anchor of the skeleton tree corresponding

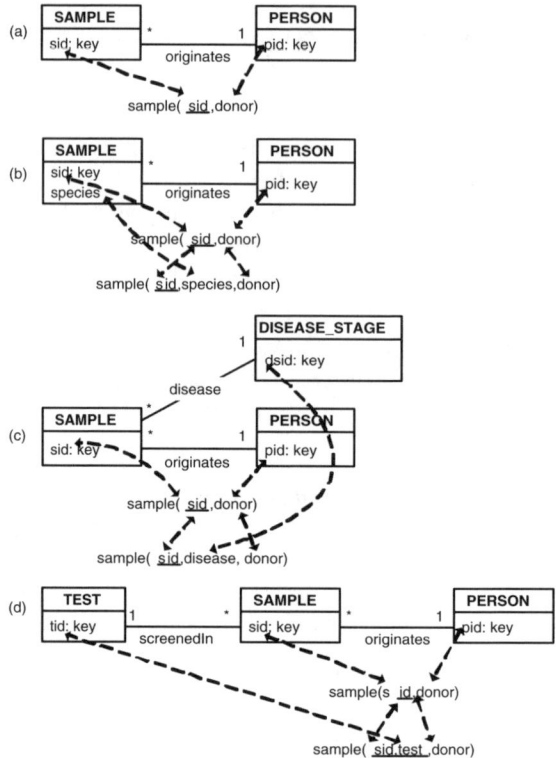

Fig. 5. Add Element to Schema

to the key as shown in Figure 5 (b) and update the semantic mapping formula to:

sample(sid, $species$, $donor$) $\leftrightarrow$
    SAMPLE(x_1), sid(x_1, sid), PERSON(x_2),
    species(x_1, $species$), originates(x_1, x_2), pid(x_2, $donor$).

(2) Add a column that is a f.k.. For example, a new column disease was added to the table sample where disease is a f.k. referencing to the key of a table T' for the concept DISEASE_STAGE. In this case, the algorithm finds a functional path from the anchor of the key skeleton tree for the key of the table sample to the anchor of the skeleton tree for the key of table T' as shown in Figure 5 (c), and updates the semantic mapping as the following candidate formula:

sample(sid, $disease$, $donor$) $\leftrightarrow$
    SAMPLE(x_1), sid(x_1, sid), PERSON(x_2),
    DISEASE_STAGE(x_3), disease(x_1, x_3), dsid(x_3, $disease$),
    originates(x_1, x_2), pid(x_2, $donor$).

Note that there may be multiple functional paths connecting the anchor SAMPLE to the anchor DISEASE_STAGE, so the user will examine the candidate formulas to choose the expected one.

(3) Add a column that becomes part of the key of the table. For example, a new column test was added to the table sample. If test is not a f.k., then the algorithm suggests to add an attribute as part of the identifier of the anchor of the skeleton tree for the key of the table. If test is a f.k., then the algorithms recomputes the skeleton tree for the key of the table as shown in Figure 5 (d), and suggests to update the semantic mapping as the following candidate formula:

sample(sid, $test$, $donor$) $\leftrightarrow$
    SAMPLE(x_1), sid(x_1, sid), PERSON(x_2),
    TEST(x_3), screenedIn(x_1, x_3), tid(x_3, $test$),
    originates(x_1, x_2), pid(x_2, $donor$).

As in case (2), the user needs to examine all candidate formulas.

Let us now consider changes that add new element(s) to the ontology. The following changes do not affect the semantic mapping: adding a new attribute which does not become part of the identifier of concepts in the s-tree, adding a new concept, and adding a new ordinary relationship. If an attribute is added to a concept in the s-tree such that the concept is an anchor of a skeleton tree, then the algorithm suggests to update the table by adding a column as part of the key or to update a f.k. that corresponds to the new identifier of the concept. Of course, the update of the f.k. must be carried out in a cascade fashion starting with the key referenced by the f.k.. If a new identifying relationship is added for changing the anchor corresponding to the key of the table from a strong entity to a weak entity, then the algorithm suggests to update the key of the table by combining the identifier of the owner entity and the local identifier of the weak entity. The semantic mapping formula is updated accordingly. ∎

For a semantic mapping, changes that delete elements from the schema can be classified into: deleting a table, deleting an attribute that is not part of the key nor a f.k. of a table, deleting a f.k. of a table, and deleting part of the key of a table. The first three deletions result in updating a semantic mapping formula that references the deleted elements without updating the ontology. The last deletion would require updating the identifier of the associated concepts in the ontology. Changes that delete elements from the ontology would require updating the associated schema in order to maintain the validity of the semantic mapping. In general, if some changes in the ontology (or schema) cause updates in the associated schema (or ontology) in order to maintain the validity of the semantic mapping, the updates will not be carried out automatically; instead, the system will prompt the suggested updates and ask the user what next action should be: executing the update or prohibiting the changes in the ontology (or schema).

The next kinds of changes are merging/splitting elements and changing constraints in schema/ontology. In this paper, we omit discussion about merging/splitting, and we use the following example to illustrate how to maintain a semantic mapping when some constraints are changed in the associated schema and ontology.

Example 5. Suppose the following existing semantic mapping formula relate a relational table treat(<u>tid</u>,<u>sgid</u>) with an s-tree $\boxed{\text{TREATMENT}}$ ---appliesTo--- $\boxed{\text{SAMPLE\_GROUP}}$ in an ontology:

   treat(tid, $sgid$) $\leftrightarrow$
      TREATMENT(x_1), tid(x_1, tid), SAMPLE_GROUP(x_2),
      appliesTo(x_1, x_2), sgid(x_3, $sgid$).

where the key of the table is the combination of both columns tid and sgid which are identifiers of concepts TREATMENT and SAMPLE_GROUP, respectively, and the relationship
appliesTo is many-to-many.

Later, the data modeler obtained a better understanding of the application by realizing that each treatment only applies to one sample group. Consequently, s/he changed the key of the treat table from the combination of columns tid and sgid to the single column tid. Having this change in the schema, the maintenance algorithm will suggest to change the relationship appliesTo from a many-to-many relationship to a functional relationship $\boxed{\text{TREATMENT}}$ ---appliesTo->-- $\boxed{\text{SAMPLE\_GROUP}}$.

Conversely, if the database designer obtained a better understanding of the application and changed the appliesTo relationship from many-to-many to functional, then the algorithm will suggest to update the key of the table treat from the combination of tid and sgid to the single column tid.

In both cases, the semantic mapping formula does not change. ■

In summary, the basic principle of maintaining semantic mappings under schema/ontology evolution is to repair the semantic relationship between a table and an s-tree according to knowledge in existing mappings and changes. Specifically, the algorithm attempts to align the key and foreign key constraints in the table with integrity constrains in the ontology by suggesting necessary updates.

6   Conclusions

A semantic mapping between a database schema and an ontology specifies a semantic relationship between the schema and the ontology. For relational schemas, we represented the semantic mapping as a set of relationships between relational tables and s-trees in an ontology. Such a relationship can be represented in terms of a formula with precisely defined semantics. Once such a semantic mapping is established, it is important to maintain the validity of the semantic mapping when the schema or ontology evolves. Mapping maintenance is a challenging problem and it will benefit from a principled and systematic solution. Here we reported on a preliminary effort to define such a solution which will empower database designers, administrators, and integrators.

Unlike the traditional solutions to the problem of schema mapping adaptation, our solution attempts to adapt both the semantic mapping and the associated schema and ontology in order to maintain the validity of the semantic mapping. Based on the previous study on discovering semantic mappings from database schemas to ontologies, we aim at repairing the semantic relationship between a table and an s-tree by analyzing the semantics in changes to align integrity constraints in schemas and ontologies.

Future work includes developing the complete algorithm and conducting experiments for testing the performance of the solution using both synthetic and real-world semantic mapping evolution scenarios. In addition, we are interested in developing solutions to the problem of maintaining general semantic mappings.

References

1. An, Y., Borgida, A., Miller, R.J., Mylopoulos, J.: A Semantic Approach to Discovering Schema Mapping Expression. In: Proceedings of International Conference on Data Engineering (ICDE) (2007)
2. An, Y., Borgida, A., Mylopoulos, J.: Constructing Complex Semantic Mappings between XML Data and Ontologies. In: Gil, Y., Motta, E., Benjamins, V.R., Musen, M.A. (eds.) ISWC 2005. LNCS, vol. 3729, pp. 6–20. Springer, Heidelberg (2005)
3. An, Y., Borgida, A., Mylopoulos, J.: Inferring Complex Semantic Mappings between Relational Tables and Ontologies from Simple Correspondences. In: Meersman, R., Tari, Z. (eds.) OTM 2005. LNCS, vol. 3760, pp. 1152–1169. Springer, Heidelberg (2005)
4. An, Y., Borgida, A., Mylopoulos, J.: Discovering the Semantics of Relational Tables through Mappings. In: Spaccapietra, S. (ed.) Journal on Data Semantics VII. LNCS, vol. 4244, pp. 1–32. Springer, Heidelberg (2006)
5. Banerjee, J., et al.: Semantics and Implementation of Schema Evolution in Object-Oriented Databases. In: Proceedings of ACM SIGMOD (1987)
6. Benatallah, B.: A Unified Framework for Supporting Dynamic Schema Evolution in Object Databases. In: Akoka, J., Bouzeghoub, M., Comyn-Wattiau, I., Métais, E. (eds.) ER 1999. LNCS, vol. 1728, pp. 16–30. Springer, Heidelberg (1999)
7. Brien, P.M., Poulovassilis, A.: Schema evolution in heterogeneous database architectures, a schema transformation approach. In: Pidduck, A.B., Mylopoulos, J., Woo, C.C., Ozsu, M.T. (eds.) CAiSE 2002. LNCS, vol. 2348. Springer, Heidelberg (2002)
8. Calvanese, D., Giacomo, G.D., Lenzerini, M., Nardi, D., Rosati, R.: Data Integration in Data Warehousing. J. of Coop. Info. Sys. 10(3), 237–271 (2001)
9. Claypool, K.T., Jin, J., Rundensteiner, E.: SERF: Schema Evolution through an Extensible, Re-usable, and Flexible Framework. In: Proceedings of International Conference on Information and Knowledge Management (CIKM) (1998)
10. Fagin, R., Kolaitis, P., Miller, R.J., Popa, L.: Data Exchange: Semantics and Query Answering. In: Calvanese, D., Lenzerini, M., Motwani, R. (eds.) ICDT 2003. LNCS, vol. 2572, pp. 207–224. Springer, Heidelberg (2002)
11. Fan, H., Poulovassilis, A.: Schema evolution in data warehousing environments — a schema transformation-based approach. In: Atzeni, P., Chu, W., Lu, H., Zhou, S., Ling, T.-W. (eds.) ER 2004. LNCS, vol. 3288, pp. 639–653. Springer, Heidelberg (2004)

12. Ferrandina, F., Ferran, G., Meyer, T., Madec, J., Zicari, R.: Schema and Database Evolution in the O2 Object Database System. In: Proceedings of International Conference on Very Large Databases (VLDB) (1995)
13. Levy, A.Y., Srivastava, D., Kirk, T.: Data Model and Query Evaluation in Global Information Systems. J. of Intelligent Info. Sys. 5(2), 121–143 (1996)
14. McCann, R., et al.: Maveric: Mapping Maintenance for Data Integration Systems. In: Proceedings of International Conference on Very Large Databases (VLDB) (2005)
15. Velegrakis, Y., Miller, R.J., Popa, L.: Mapping Adapdation under Evolving Schemas. In: Proceedings of International Conference on Very Large Databases (VLDB) (2003)
16. Yu, C., Popa, L.: Semantic Adaptation of Schema Mappings when Schema Evolve. In: Proceedings of International Conference on Very Large Databases (VLDB) (2005)

Author Index

Printing: Mercedes-Druck, Berlin
Binding: Stein+Lehmann, Berlin

Lecture Notes in Computer Science

Sublibrary 3: Information Systems and Application, incl. Internet/Web and HCI

For information about Vols. 1– 4717
please contact your bookseller or Springer

Vol. 5120: S. Helal, S. Mitra, J. Wong, C.K. Chang, M. Mokhtari (Eds.), Smart Homes and Health Telematics. XV, 220 pages. 2008.

Vol. 5105: K. Miesenberger, J. Klaus, W. Zagler, A. Karshmer (Eds.), Computers Helping People with Special Needs. XXVIII, 1350 pages. 2008.

Vol. 5094: V. Atluri (Ed.), Data and Applications Security XXII. IX, 347 pages. 2008.

Vol. 5093: Z. Pan, X. Zhang, A. El Rhalibi, W. Woo, Y. Li (Eds.), Technologies for E-Learning and Digital Entertainment. XVII, 791 pages. 2008.

Vol. 5080: Z. Pan, A.D. Cheok, W. Müller, A. El Rhalibi (Eds.), Transactions on Edutainment I. X, 305 pages. 2008.

Vol. 5075: C.C. Yang, H. Chen, M. Chau, K. Chang, S.-D. Lang, P.S. Chen, R. Hsieh, D. Zeng, F.-Y. Wang, K.M. Carley, W. Mao, J. Zhan (Eds.), Intelligence and Security Informatics. XXII, 522 pages. 2008.

Vol. 5074: Z. Bellahsène, M. Léonard (Eds.), Advanced Information Systems Engineering. XVII, 588 pages. 2008.

Vol. 5071: A. Gray, K. Jeffery, J. Shao (Eds.), Sharing Data, Information and Knowledge. XI, 293 pages. 2008.

Vol. 5069: B. Ludäscher, N. Mamoulis (Eds.), Scientific and Statistical Database Management. XIII, 620 pages. 2008.

Vol. 5068: S. Lee, H. Choo, S. Ha, I.C. Shin (Eds.), Computer Human Interaction. XVII, 458 pages. 2008.

Vol. 5066: M. Tscheligi, M. Obrist, A. Lugmayr (Eds.), Changing Television Environments. XV, 324 pages. 2008.

Vol. 5061: F.E. Sandnes, Y. Zhang, C. Rong, L.T. Yang, J. Ma (Eds.), Ubiquitous Intelligence and Computing. XVI, 763 pages. 2008.

Vol. 5053: R. Meier, S. Terzis (Eds.), Distributed Applications and Interoperable Systems. XI, 303 pages. 2008.

Vol. 5039: E. Kapetanios, V. Sugumaran, M. Spiliopoulou (Eds.), Natural Language and Information Systems. XIX, 386 pages. 2008.

Vol. 5034: R. Fleischer, J. Xu (Eds.), Algorithmic Aspects in Information and Management. XI, 350 pages. 2008.

Vol. 5033: H. Oinas-Kukkonen, P. Hasle, M. Harjumaa, K. Segerståhl, P. Øhrstrøm (Eds.), Persuasive Technology. XIV, 287 pages. 2008.

Vol. 5024: M. Ferre (Ed.), Haptics: Perception, Devices and Scenarios. XXIII, 950 pages. 2008.

Vol. 5021: S. Bechhofer, M. Hauswirth, J. Hoffmann, M. Koubarakis (Eds.), The Semantic Web: Research and Applications. XIX, 897 pages. 2008.

Vol. 5017: T. Nanya, F. Maruyama, A. Pataricza, M. Malek (Eds.), Service Availability. XII, 225 pages. 2008.

Vol. 5013: J. Indulska, D.J. Patterson, T. Rodden, M. Ott (Eds.), Pervasive Computing. XIV, 315 pages. 2008.

Vol. 5006: R. Kowalczyk, M. Huhns, M. Klusch, Z. Maamar, Q.B. Vo (Eds.), Service-Oriented Computing: Agents, Semantics, and Engineering. X, 154 pages. 2008.

Vol. 5005: V. Christophides, M. Collard, C. Gutierrez (Eds.), Semantic Web, Ontologies and Databases. VII, 153 pages. 2008.

Vol. 4997: B. Monien, U.-P. Schroeder (Eds.), Algorithmic Game Theory. XI, 363 pages. 2008.

Vol. 4993: H. Li, T. Liu, W.-Y. Ma, T. Sakai, K.-F. Wong, G. Zhou (Eds.), Information Retrieval Technology. XIII, 685 pages. 2008.

Vol. 4976: Y. Zhang, G. Yu, E. Bertino, G. Xu (Eds.), Progress in WWW Research and Development. XVIII, 699 pages. 2008.

Vol. 4956: C. Macdonald, I. Ounis, V. Plachouras, I. Ruthven, R.W. White (Eds.), Advances in Information Retrieval. XXI, 719 pages. 2008.

Vol. 4952: C. Floerkemeier, M. Langheinrich, E. Fleisch, F. Mattern, S.E. Sarma (Eds.), The Internet of Things. XIII, 378 pages. 2008.

Vol. 4947: J.R. Haritsa, R. Kotagiri, V. Pudi (Eds.), Database Systems for Advanced Applications. XXII, 713 pages. 2008.

Vol. 4936: W. Aiello, A. Broder, J. Janssen, E.E. Milios (Eds.), Algorithms and Models for the Web-Graph. X, 167 pages. 2008.

Vol. 4932: S. Hartmann, G. Kern-Isberner (Eds.), Foundations of Information and Knowledge Systems. XII, 397 pages. 2008.

Vol. 4928: A.H.M. ter Hofstede, B. Benatallah, H.-Y. Paik (Eds.), Business Process Management Workshops. XIII, 518 pages. 2008.

Vol. 4918: N. Boujemaa, M. Detyniecki, A. Nürnberger (Eds.), Adaptive Multimedial Retrieval: Retrieval, User, and Semantics. XI, 265 pages. 2008.

Vol. 4903: S. Satoh, F. Nack, M. Etoh (Eds.), Advances in Multimedia Modeling. XIX, 510 pages. 2008.

Vol. 4900: S. Spaccapietra (Ed.), Journal on Data Semantics X. XIII, 265 pages. 2008.

Vol. 4892: A. Popescu-Belis, S. Renals, H. Bourlard (Eds.), Machine Learning for Multimodal Interaction. XI, 308 pages. 2008.

Vol. 4882: T. Janowski, H. Mohanty (Eds.), Distributed Computing and Internet Technology. XIII, 346 pages. 2007.

Vol. 4881: H. Yin, P. Tino, E. Corchado, W. Byrne, X. Yao (Eds.), Intelligent Data Engineering and Automated Learning - IDEAL 2007. XX, 1174 pages. 2007.

Vol. 4877: C. Thanos, F. Borri, L. Candela (Eds.), Digital Libraries: Research and Development. XII, 350 pages. 2007.

Vol. 4872: D. Mery, L. Rueda (Eds.), Advances in Image and Video Technology. XXI, 961 pages. 2007.

Vol. 4871: M. Cavazza, S. Donikian (Eds.), Virtual Storytelling. XIII, 219 pages. 2007.

Vol. 4858: X. Deng, F.C. Graham (Eds.), Internet and Network Economics. XVI, 598 pages. 2007.

Vol. 4857: J.M. Ware, G.E. Taylor (Eds.), Web and Wireless Geographical Information Systems. XI, 293 pages. 2007.

Vol. 4853: F. Fonseca, M.A. Rodríguez, S. Levashkin (Eds.), GeoSpatial Semantics. X, 289 pages. 2007.

Vol. 4836: H. Ichikawa, W.-D. Cho, I. Satoh, H.Y. Youn (Eds.), Ubiquitous Computing Systems. XIII, 307 pages. 2007.

Vol. 4832: M. Weske, M.-S. Hacid, C. Godart (Eds.), Web Information Systems Engineering – WISE 2007 Workshops. XV, 518 pages. 2007.

Vol. 4831: B. Benatallah, F. Casati, D. Georgakopoulos, C. Bartolini, W. Sadiq, C. Godart (Eds.), Web Information Systems Engineering – WISE 2007. XVI, 675 pages. 2007.

Vol. 4825: K. Aberer, K.-S. Choi, N. Noy, D. Allemang, K.-I. Lee, L. Nixon, J. Golbeck, P. Mika, D. Maynard, R. Mizoguchi, G. Schreiber, P. Cudré-Mauroux (Eds.), The Semantic Web. XXVII, 973 pages. 2007.

Vol. 4823: H. Leung, F. Li, R. Lau, Q. Li (Eds.), Advances in Web Based Learning – ICWL 2007. XIV, 654 pages. 2008.

Vol. 4822: D.H.-L. Goh, T.H. Cao, I.T. Sølvberg, E. Rasmussen (Eds.), Asian Digital Libraries. XVII, 519 pages. 2007.

Vol. 4820: T.G. Wyeld, S. Kenderdine, M. Docherty (Eds.), Virtual Systems and Multimedia. XII, 215 pages. 2008.

Vol. 4816: B. Falcidieno, M. Spagnuolo, Y. Avrithis, I. Kompatsiaris, P. Buitelaar (Eds.), Semantic Multimedia. XII, 306 pages. 2007.

Vol. 4813: I. Oakley, S.A. Brewster (Eds.), Haptic and Audio Interaction Design. XIV, 145 pages. 2007.

Vol. 4810: H.H.-S. Ip, O.C. Au, H. Leung, M.-T. Sun, W.-Y. Ma, S.-M. Hu (Eds.), Advances in Multimedia Information Processing – PCM 2007. XXI, 834 pages. 2007.

Vol. 4809: M.K. Denko, C.-s. Shih, K.-C. Li, S.-L. Tsao, Q.-A. Zeng, S.H. Park, Y.-B. Ko, S.-H. Hung, J.-H. Park (Eds.), Emerging Directions in Embedded and Ubiquitous Computing. XXXV, 823 pages. 2007.

Vol. 4808: T.-W. Kuo, E. Sha, M. Guo, L.T. Yang, Z. Shao (Eds.), Embedded and Ubiquitous Computing. XXI, 769 pages. 2007.

Vol. 4806: R. Meersman, Z. Tari, P. Herrero (Eds.), On the Move to Meaningful Internet Systems 2007: OTM 2007 Workshops, Part II. XXXIV, 611 pages. 2007.

Vol. 4805: R. Meersman, Z. Tari, P. Herrero (Eds.), On the Move to Meaningful Internet Systems 2007: OTM 2007 Workshops, Part I. XXXIV, 757 pages. 2007.

Vol. 4804: R. Meersman, Z. Tari (Eds.), On the Move to Meaningful Internet Systems 2007: CoopIS, DOA, ODBASE, GADA, and IS, Part II. XXIX, 683 pages. 2007.

Vol. 4803: R. Meersman, Z. Tari (Eds.), On the Move to Meaningful Internet Systems 2007: CoopIS, DOA, ODBASE, GADA, and IS, Part I. XXIX, 1173 pages. 2007.

Vol. 4802: J.-L. Hainaut, E.A. Rundensteiner, M. Kirchberg, M. Bertolotto, M. Brochhausen, Y.-P.P. Chen, S.S.-S. Cherfi, M. Doerr, H. Han, S. Hartmann, J. Parsons, G. Poels, C. Rolland, J. Trujillo, E. Yu, E. Zimányie (Eds.), Advances in Conceptual Modeling – Foundations and Applications. XIX, 420 pages. 2007.

Vol. 4801: C. Parent, K.-D. Schewe, V.C. Storey, B. Thalheim (Eds.), Conceptual Modeling - ER 2007. XVI, 616 pages. 2007.

Vol. 4797: M. Arenas, M.I. Schwartzbach (Eds.), Database Programming Languages. VIII, 261 pages. 2007.

Vol. 4796: M. Lew, N. Sebe, T.S. Huang, E.M. Bakker (Eds.), Human–Computer Interaction. X, 157 pages. 2007.

Vol. 4794: B. Schiele, A.K. Dey, H. Gellersen, B. de Ruyter, M. Tscheligi, R. Wichert, E. Aarts, A. Buchmann (Eds.), Ambient Intelligence. XV, 375 pages. 2007.

Vol. 4777: S. Bhalla (Ed.), Databases in Networked Information Systems. X, 329 pages. 2007.

Vol. 4761: R. Obermaisser, Y. Nah, P. Puschner, F.J. Rammig (Eds.), Software Technologies for Embedded and Ubiquitous Systems. XIV, 563 pages. 2007.

Vol. 4747: S. Džeroski, J. Struyf (Eds.), Knowledge Discovery in Inductive Databases. X, 301 pages. 2007.

Vol. 4744: Y. de Kort, W. IJsselsteijn, C. Midden, B. Eggen, B.J. Fogg (Eds.), Persuasive Technology. XIV, 316 pages. 2007.

Vol. 4740: L. Ma, M. Rauterberg, R. Nakatsu (Eds.), Entertainment Computing – ICEC 2007. XXX, 480 pages. 2007.

Vol. 4730: C. Peters, P. Clough, F.C. Gey, J. Karlgren, B. Magnini, D.W. Oard, M. de Rijke, M. Stempfhuber (Eds.), Evaluation of Multilingual and Multi-modal Information Retrieval. XXIV, 998 pages. 2007.

Vol. 4723: M. R. Berthold, J. Shawe-Taylor, N. Lavrač (Eds.), Advances in Intelligent Data Analysis VII. XIV, 380 pages. 2007.

Vol. 4721: W. Jonker, M. Petković (Eds.), Secure Data Management. X, 213 pages. 2007.

Vol. 4718: J. Hightower, B. Schiele, T. Strang (Eds.), Location- and Context-Awareness. X, 297 pages. 2007.